Thomas Neusius

Thermal Fluctuations of Biomolecules

Thomas Neusius

Thermal Fluctuations of Biomolecules

An Approach to Understand the Subdiffusion in the Internal Dynamics of Peptides and Proteins

Südwestdeutscher Verlag für Hochschulschriften

Impressum / Imprint
Bibliografische Information der Deutschen Nationalbibliothek: Die Deutsche Nationalbibliothek verzeichnet diese Publikation in der Deutschen Nationalbibliografie; detaillierte bibliografische Daten sind im Internet über http://dnb.d-nb.de abrufbar.
Alle in diesem Buch genannten Marken und Produktnamen unterliegen warenzeichen-, marken- oder patentrechtlichem Schutz bzw. sind Warenzeichen oder eingetragene Warenzeichen der jeweiligen Inhaber. Die Wiedergabe von Marken, Produktnamen, Gebrauchsnamen, Handelsnamen, Warenbezeichnungen u.s.w. in diesem Werk berechtigt auch ohne besondere Kennzeichnung nicht zu der Annahme, dass solche Namen im Sinne der Warenzeichen- und Markenschutzgesetzgebung als frei zu betrachten wären und daher von jedermann benutzt werden dürften.

Bibliographic information published by the Deutsche Nationalbibliothek: The Deutsche Nationalbibliothek lists this publication in the Deutsche Nationalbibliografie; detailed bibliographic data are available in the Internet at http://dnb.d-nb.de.
Any brand names and product names mentioned in this book are subject to trademark, brand or patent protection and are trademarks or registered trademarks of their respective holders. The use of brand names, product names, common names, trade names, product descriptions etc. even without a particular marking in this work is in no way to be construed to mean that such names may be regarded as unrestricted in respect of trademark and brand protection legislation and could thus be used by anyone.

Verlag / Publisher:
Südwestdeutscher Verlag für Hochschulschriften
ist ein Imprint der / is a trademark of
OmniScriptum GmbH & Co. KG
Heinrich-Böcking-Str. 6-8, 66121 Saarbrücken, Deutschland / Germany
Email: info@svh-verlag.de

Herstellung: siehe letzte Seite /
Printed at: see last page
ISBN: 978-3-8381-1145-2

Zugl. / Approved by: Heidelberg, Ruprecht-Karls-Universität, Dissertation, 2009

Copyright © 2009 OmniScriptum GmbH & Co. KG
Alle Rechte vorbehalten. / All rights reserved. Saarbrücken 2009

Zusammenfassung

Auf der Basis von Molekulardynamik-Simulationen werden die thermischen Fluktuationen in den inneren Freiheitsgraden von Biomolekülen, wie Oligopeptiden oder einer β-Schleife, untersucht, unter besonderer Beachtung der zeitgemittelten, mittleren quadratischen Verschiebung (MSD). Die Simulationen lassen in einem Bereich von 10^{-12} bis 10^{-8} s Subdiffusion im thermischen Gleichgewicht erkennen. Mögliche Modelle subdiffusiver Fluktuationen werden vorgestellt und diskutiert. Der zeitkontinuierliche Zufallslauf (CTRW), dessen Wartezeitverteilung einem Potenzgesetz folgt, wird als mögliches Modell untersucht. Während das ensemble-gemittelte MSD eines freien CTRW Subdiffusion zeigt, ist dies beim zeitgemittelten MSD nicht der Fall. Hier wird gezeigt, daß der CTRW in einem begrenzten Volumen ein zeitgemitteltes MSD aufweist, das ab einer kritischen Zeit subdiffusiv ist. Eine analytische Näherung des zeitgemittelten MSD wird hergeleitet und mit Computersimulationen von CTRW-Prozessen verglichen. Ein Vergleich der Parameter des begrenzten CTRW mit den Ergebnissen der Molekulardynamik-Simulationen zeigt, daß CTRW die Subdiffusion der inneren Freiheitgrade nicht erklären kann. Eher muß die Subdiffusion als Konsequenz der fraktalartigen Struktur des zugänglichen Volumens im Konfigurationsraum betrachtet werden. Mit Hilfe von Übergangsmatrizen kann der Konfigurationsraum durch ein Netzwerkmodell angenähert werden. Die Hausdorff-Dimension der Netzwerke liegt im fraktalen Bereich. Die Netzwerkmodelle erlauben eine gute Modellierung der Kinetik auf Zeitskalen ab 100 ps.

Abstract

Molecular dynamics (MD) simulations are used to analyze the thermal fluctuations in the internal coordinates of biomolecules, such as oligopeptide chains or a β-hairpin, with a focus on the time-averaged mean squared displacement (MSD). The simulations reveal configurational subdiffusion at equilibrium in the range from 10^{-12} to 10^{-8} s. Potential models of subdiffusive fluctuations are discussed. Continuous time random walks (CTRW) with a power-law distribution of waiting times are examined as a potential model of subdiffusion. Whereas the ensemble-averaged MSD of an unbounded CTRW exhibits subdiffusion, the time-averaged MSD does not. Here, we demonstrate that in a bounded (confined) CTRW the time-averaged MSD indeed exhibits subdiffusion beyond a critical time. An analytical approximation to the time-averaged MSD is derived and compared to a numerical MSD obtained from simulations of the model. Comparing the parameters of the confined CTRW to the results obtained from the MD trajectories, the CTRW is disqualified as a model of the subdiffusive internal fluctuations. Subdiffusion arises rather from the fractal-like structure of the accessible configuration space. Employing transition matrices, the configuration space is represented by a network, the Hausdorff dimensions of which are found to be fractal. The network representation allows the kinetics to be correctly reproduced on time scales of 100 ps and above.

Meinen Eltern.

Few people who are not actually practitioners of a mature science realize how much mop-up work of this sort a paradigm leaves to be done or quite how fascinating such work can prove in the execution. And these points need to be understood. Mopping-up operations are what engage most scientists throughout their careers. They constitute what I am here calling normal science.

THOMAS S. KUHN

VORWORT

Die vorliegende Arbeit – wie sollte es anders sein – konnte nur gelingen Dank der Unterstützung und Hilfe, die ich in den letzten gut vier Jahren erfahren habe. An erster Stelle ist mein Betreuer Prof. Jeremy C. Smith zu nennen, der mir die Möglichkeit gegeben hat, mich mit einem spannenden Thema zu beschäftigen, meine eigenen Ansätze mit großem Freiraum zu entwickeln und immer weiter zu vertiefen. Insbesondere seine Geduld bei bei Erstellung der Manuskripte für die geplanten Veröffentlichungen war bemerkenswert.

Weiterhin danke ich Prof. Igor M. Sokolov von der Humboldt-Universität zu Berlin, der mir bedeutende Anregungen geben hat und mich früh bestärkte, das Thema weiterzuverfolgen – eine wichtige Motivation zum richtigen Zeitpunkt.

Isabella Daidone hat mir freundlicherweise ihre umfangreichen MD-Simulationen zur Verfügung gestellt und mir bei vielen Detailfragen weitergeholfen. Diskussionen mit ihr, sowie mit Frank Noé und Dieter Krachtus waren wesentlicher Anlaß, das Thema immer wieder von verschiedenen Seiten zu betrachten. Nadia Elgobashi-Meinhardt hat mit ihren Korrekturen des Dissertationsmanuskripts wesentlich dazu beigetragen, daß der Text sich einer lesbaren Form zumindest angenähert hat. Lars Meinhold stand mir immer wieder mit Rat und Tat zur Seite. Ein besonderer Dank geht an Thomas Splettstößer. Seine spontane Quartierwahl im winterlichen Knoxville hat mich vermutlich vor Depressionen bewahrt.

Bei allen Mitarbeitern der CMB-Gruppe am Interdisziplinären Zentrum für wissenschaftliches Rechnen möchte ich mich für die gute Zusammenarbeit und die Unterstützung bei Problemen bedanken. Ellen Vogel war eine unverzichtbare Hilfe bei allen formalen Dingen, insbesondere beim Thema Dienstreisen. Bogdan Costescu hat die Rechner der Arbeitsgruppe am Laufen gehalten; ohne ihn wären wir schon lange aufgeschmissen.

Ohne die Unterstützung meiner Eltern wäre ein Studium in Heidelberg kaum möglich gewesen. Schließlich möchte ich mich bei meiner Frau Chrisi für ihr Verständnis für meine längeren Abwesenheiten und die moralische Stärkung während der Durststrecken bedanken, die wohl jedes Promotionsprojekt mit sich bringt.

Oak Ridge, Tennessee,
im März 2009 *Thomas Neusius*

Contents

Zusammenfassung   ii

Abstract   iii

Vorwort   vii

1 Introduction   1
    1.1 The dynamics of biomolecules 3
    1.2 Thesis outline 6

2 Thermal Fluctuations   9
    2.1 Very brief history of the kinetic theory 9
    2.2 Basic concepts 11
    2.3 Brownian motion 16
    2.4 From random walks to diffusion 22
    2.5 Anomalous diffusion 25

3 Molecular Dynamics Simulations of Biomolecules   29
    3.1 Numerical integration 32
    3.2 Force field 34
    3.3 The energy landscape 42
    3.4 Normal Mode Analysis (NMA) 44
    3.5 Principal Component Analysis (PCA) 46
    3.6 Convergence 48

4 Study of Biomolecular Dynamics — 51
4.1 Setup of the MD simulations 51
4.2 Thermal fluctuations in biomolecules 53
4.3 Thermal kinetics of biomolecules 65
4.4 Conclusion 73

5 Modeling Subdiffusion — 75
5.1 Zwanzig's projection formalism 75
5.2 Chain dynamics 79
5.3 The Continuous Time Random Walk 83
  5.3.1 Ensemble averages and CTRW 87
  5.3.2 Time averages and CTRW 90
  5.3.3 Confined CTRW 95
  5.3.4 Application of CTRW to internal biomolecular dynamics . 98
5.4 Diffusion on networks 102
  5.4.1 From the energy landscape to transition networks 103
  5.4.2 Networks as Markov chain models 104
  5.4.3 Subdiffusion in fractal networks 106
  5.4.4 Transition networks as models of the configuration space . 113
  5.4.5 Eigenvector-decomposition and diffusion in networks . . . 120
5.5 Conclusion 126

6 Concluding Remarks and Outlook — 129
6.1 Conclusion 129
6.2 Outlook 133

References — 135

Abbreviations and Symbols — 147

Index — 149

Appendix — 153
A Diffusion equation solved by Fourier decomposition 153
B Constrained Verlet algorithm – LINCS 155

C	Derivation of the Rouse ACF		157
D	Fractional diffusion equation		159
	D.1	Gamma function	159
	D.2	Fractional derivatives	160
	D.3	The fractional diffusion equation	161
	D.4	The Mittag-Leffler function	163
E	Exponentials and power laws		164

Zitate in den Kapitelüberschriften   **167**

> Jedoch das Gebiet, welches der unbedingten Herrschaft der vollendeten Wissenschaft unterworfen werden kann, ist leider sehr eng, und schon die organische Welt entzieht sich ihm größtenteils.
> HERMANN VON HELMHOLTZ

CHAPTER 1

INTRODUCTION

Some of the most fruitful physical concepts were motivated by biological or medical observations. In particular, thermodynamics owe some of their fundamental insights to the physiological interests of their pioneers. Julius Robert von Mayer was a medical doctor by training and the first to realize that heat is nothing else than a form of energy. Also Hermann von Helmholtz's interest in thermodynamics was closely connected to his physiological studies.

The discovery of the motion of pollen grains by the botanist Robert Brown in 1827 [1] and the development of the diffusion equation by the physiologist Adolf Fick in 1855 [2] are fundamental landmarks for the understanding of transport phenomena. In the framework of the kinetic theory of heat, Albert Einstein could demonstrate in 1905 [3] that the atomistic structure of matter brings about thermal fluctuations, *i.e.*, the irregular, chaotic trembling of atoms and molecules. The thermal fluctuations give rise to the zig-zag motion, *e.g.* of pollen grains or other small particles suspended in a fluid. In 1908, Jean Perrin's skillful experimental observations revealed that Brownian motion is to be understood as a molecular fluctuation phenomenon, as suggested by Einstein [4]. Brownian motion is an extremely fundamental process, and it is still a very active field of research [5].

Thermal fluctuations are microscopic effects. However, what is observed on the macroscopic level are thermodynamic quantities like, *e.g.*, temperature, pressure, and heat capacity. Statistical physics, as developed by Gibbs and Boltzmann, based the thermodynamic quantities on microscopic processes and derived the equations of thermodynamics as limiting cases for infinitely many particles. In so doing, Fick's diffusion equation can be obtained from Brownian motion. Thus, diffusion processes are powered by thermal energy.

The diffusion equation turned out to be an approximate description of a fairly large class of microscopic and other processes, in which randomness has a critical influence, *e.g.*, osmosis, the spontaneous mixing of gases, the charge current in metals, or disease spreading. In particular, diffusion is an important biological transport mechanism on the cellular level.

Among the microscopic processes giving rise to diffusion, one of the most simple scenarios is the random walk model. Due to its simplicity, the random walk proves to be applicable in a wide range of different situations. Whenever microscopic processes can be described reasonably with random walk models, the diffusion equation can be employed for the modeling of the collective dynamics of a large number of such processes.

Despite the fruitfulness of the diffusion equation, deviations from the classical diffusion became apparent as experimental skills advanced. The need to broaden the concept of diffusion led to the development of the *fractional diffusion equation* [6–8] and the corresponding microscopic processes like, *e.g.*, the *continuous time random walk* (CTRW) [8, 9]. The wider approach paved the way to apply stochastic models to an even larger variety of situations in which randomness is present.

A major challenge of contemporary science is the understanding of the behavior of proteins and peptides. Those biomolecules are among the smallest building blocks of life[1]. They are employed by the cells for a plethora of functions, *e.g.* metabolism, signal transduction, transport or mechanical work. Proteins and peptides are chains of amino acids. Proteins usually form a stable conformation, but this does not mean that they are static objects. As proteins (and peptides) are molecular objects they are directly influenced by the presence of thermal fluctuations; their precise structure undergoes dynamical changes due to strokes and kicks from the surrounding molecules [10]. Hence, any description of the internal

---

[1] The name *protein* – $\pi\rho\tilde{\omega}\tau\varepsilon\iota o\varsigma$ (proteios) means in ancient Greek „I take the first place" – emphasizes its central role for life. The denomination had been introduced by the Swedish chemist Jöns Jakob Berzelius in 1838.

dynamics of proteins and peptides must take into account the influence of thermal agitation, a task that can only be accomplished in statistical terms.

The thesis at hand presents one small step towards an understanding of the principles that govern internal dynamics of peptides and proteins. The viewpoint taken up in this work is to look at the most general features of the thermal fluctuations seen in peptides and proteins. More precisely, this work is focused on the long time correlations in the dynamics of biomolecules. In particular, the decelerated time dependence of the mean squared displacement (MSD) is studied in some detail – a phenomenon referred to as subdiffusivity.

A thorough understanding of the possible stochastic models is required in order to assess their fruitfulness in the context of biomolecular, thermal motion. On that account, this work dwells mainly on the stochastic models, some of which may be of interest not only in biophysics. The theoretical considerations are compared to molecular dynamics (MD) simulations of peptides and proteins.

One of the main result of this thesis is that the structure of the accessible volume in configurational space is responsible for the subdiffusive MSD of the biomolecule rather than the distribution of barrier heights or the local minima's effective depths. Instead of trap models or CTRW, which rely on the statistics of trap depths or waiting times, respectively, and ignore the geometry of the configuration space, we suggest network models of the configuration space.

1.1 The dynamics of biomolecules

Peptides and proteins are built from a set of twenty-two amino acids (called *residues*) [11]. All amino acids contain a carboxyl group and an amino group. Two amino acids can form a chemical bond, the *peptide bond*, by releasing a water molecule and coupling the carboxyl group of one residue to the amino group of an adjacent residue. In that way, chain-like molecules can be built with up to thousands of residues[2]. The chain of peptide bonds is referred to as the *backbone* of the molecule. The cell synthesizes the sequence of a protein (primary structure)

---

[2] Most proteins have some hundreds of residues. The largest known polypeptide is titin, with more than twenty thousand residues [12, 13].

which is encoded in the deoxyribonucleic acid (DNA). In aqueous solution proteins spontaneously assume a compact spacial structure; they *fold* into the *natural conformation* or *native state* (secondary structure). The natural conformation is determined by the sequence of the protein and stabilized by the formation of hydrogen bonds. It is an object of current research to discover how the molecule is able to find its natural conformation as quickly as it does in nature and in experiment, the problem of protein folding.

Peptides are smaller than proteins and usually they do not fold into a fixed three dimensional structure. The number of residues needed to make up a protein varies, but the commonly used terminology speaks of proteins starting from 20 to 30 residues.

Experimental techniques that allow the observation of biomolecules on an atomic level started being developed in the 1960s. The first protein structures to be resolved by X-ray crystallography were myoglobin [14] and hemoglobin [15]. For their achievement the Nobel Prize was awarded to John Kendrew and Max Perutz in 1962. X-ray crystallography, small-angle X-ray scattering and nuclear magnetic resonance (NMR) techniques are used for structure determination. Dynamical information is provided, *e.g.*, by NMR, infrared spectroscopy, Raman spectroscopy and quasi-elastic neutron scattering. Fluorescence methods give access to single molecule dynamics [10].

Besides experiment and theory a third approach has been established in the field of the dynamics of biomolecules. Computer simulation is a widespread tool for exploring the behavior of molecules. Computational methods allow mathematical models to be exploited, the complexity of which does not permit analytical solutions. Computer simulations help to bridge the gap between theory and experiment. In particular, MD simulations are useful with respect to dynamics: they produce a time series of coordinates of the molecule investigated, the so-called *trajectory*, by a numerical integration of the Newtonian equations of motion. The numerical integration gives access to the detailed time behavior [16–18].

The conformation of proteins is a prerequisite for the understanding of the biological function for which the molecule is designed. However, the protein is not frozen in the natural conformation. Rather the conformation is the ensemble

1.1 The dynamics of biomolecules

of several, nearly isoenergetic conformational substates [19–22]. Thus, at ambient temperature, the protein maintains a flexibility that is necessary for its biological function [10]. The subconformations can be seen as local minima on the high-dimensional complex energy landscape. The substates are separated by energetic barriers, a crossing of which occurs due to thermal activation [23]. There are further classifications of the substates into a hierarchical organization [10, 20, 22, 24]. The properties of the energy landscape determine the dynamical behavior of the molecule.

The concept of an energy landscape was developed in the context of the dynamics of glass formers [25]. Indeed, the dynamics of proteins exhibit some similarities with the dynamics of glasses [20, 26], *e.g.* non-exponential relaxation patterns [27–29] and non-Arrhenius-like temperature dependence [29, 30]. At a temperature $T_g \approx 200$ K a transition of the dynamical behavior is observed for many proteins; below T_g the amplitude of the thermal fluctuations increases proportional to the temperature. Above T_g, a sharp increase of the overall fluctuations is revealed by quasi-elastic neutron scattering [31–34], X-ray scattering [35, 36], and Mößbauer spectroscopy [37–41]. Water plays a dominant role in the dynamical transition, also sometimes referred to as *glass transition* [29, 34, 42–45]. It is controversial whether the protein ceases to function below T_g [36, 46, 47]. Recently, terahertz time domain spectroscopy revealed the dynamical transition of poly-alanine, a peptide without secondary structure [48].

The glass-like behavior of proteins is seen in a variety of different molecules, such as myoglobin, lysozyme, tRNA, and polyalanine [31, 33, 34, 48]. The presence of glassy dynamics in very different systems indicates that various peptides and proteins, albeit different in size and shape, share similar dynamical features [48].

Due to the presence of strong thermal fluctuations in the environment of biomolecules, theoretical attempts to model the dynamics have to include stochastic contributions. The main focus of this work is on *subdiffusion* in the internal dynamics, *i.e.*, a behavior that is characterized by a MSD that exhibits a time dependence of $\sim t^\alpha$ (where $0 < \alpha < 1$) rather than a linear time dependence. This non-linear behavior is seen in a variety of experiments and was also found in MD simulations [49]:

- The rebinding of carbon monoxide and myoglobin upon flash dissociation follows a stretched exponential pattern [23]. The non-exponential relaxation indicates fractional dynamics [50].

- In single-molecule experiments, the fluorescence lifetime of a flavin quenched by electron transfer from a tyrosine residue allows the distance fluctuations between the two residues to be measured. The statistics of the distance fluctuations has revealed long-lasting autocorrelations corresponding to subdiffusive dynamics [27, 28, 51, 52].

- Spin echo neutron scattering experiments on lysozyme reveal long-lasting correlations attributed to fractional dynamics [53, 54].

- In MD simulations, lysozyme in aqueous solution exhibits subdiffusive behavior [53].

There have been different approaches to describing subdiffusive dynamics and to understanding the mechanism by which it emerges.

1.2 Thesis outline

This thesis is organized as follows. The fundamental concept of diffusion theory, its connection to kinetic theory, Brownian motion, and random walk models are reviewed in Chap. 2. The methodology of MD simulations and common analysis tools are assembled in Chap. 3. The sections in Chap. 3 cover the integration algorithm, the terms included in the interaction potential, the energy landscape picture, and the techniques of normal mode analysis and principal component analysis. Sec. 3.6 briefly comments on the problem of convergence.

In Chap. 4, we present the MD simulations performed for a β-hairpin molecule folding into a β-turn and oligopeptide chains lacking a specific native state. After a description of the simulation set up in Sec. 4.1, the thermal fluctuations seen in the MD simulations are analyzed with respect to the principal components. The potentials of mean force are obtained and characterized with respect to delocalization and anharmonicity. Some of the different techniques employed for

1.2 Thesis outline

the modeling of the water dynamics in MD simulations, including explicit water, Langevin thermostat, and the generalized Born model, are compared in Sec. 4.2. In particular, the accuracy of the implicit water simulations with respect to the kinetic aspects is assessed by a comparison to the explicit water simulation of the same system. Internal fluctuations of biomolecules have been reported to be subdiffusive; the claim is based on experimental and MD simulation results. Therefore, we address the following question:

> *Is subdiffusivity a general feature of internal fluctuations in biomolecules?*

The kinetic aspects of the MD trajectories are discussed with particular emphasis on the MSD, which is found to be subdiffusive. The next chapter, Chap. 5, is dedicated to the central question of the present thesis.

> *What is the mechanism that causes the internal dynamics of biomolecules to be subdiffusive?*

The chapter is subdivided in five sections. The first section reviews briefly the projection operator approach, as developed by R. Zwanzig and H. Mori in the 1960s. The projection operator approach describes the dynamics with a reduced set of dynamical coordinates, considered to be relevant. It is demonstrated by Zwanzig's approach that correlations emerge from the ignorance against dynamical coordinates deemed irrelevant. As a consequence, the dynamics appear to exhibit time correlations and to be non-Markovian. Sec. 5.2 deals with the Rouse chain polymer model as an example of a harmonic chain which exhibits subdiffusive distance fluctuations. Trap models, prominently the CTRW model, are examined in detail in Sec. 5.3. As the CTRW model has been discussed as a candidate for modeling subdiffusive, biomolecular motion, we ask:

> *Can the CTRW model exhibit subdiffusivity in the time-averaged MSD?*

We emphasize that subdiffusivity in the time-averaged MSD is a prerequisite for the application of CTRWs to MD simulations, as the trajectories provide time-averaged quantities. However, the interest in the question may be not confined to the context of biomolecular dynamics, as CTRWs are used in various, very different, physical fields. The answer to the above question prompts to the ergodicity breaking found in the CTRW with a power-law tailed distribution of waiting times. The non-ergodicity of the model requires a careful treatment of time and ensemble averaging procedures. The influence of a finite volume boundary condition is examined. The theoretical results are corroborated by extensive simulations of the CTRW model.

An alternative model of subdiffusive fluctuations is based on the network representation of the configuration space. The network approach is presented in Sec. 5.4. The network representation allows the time evolution of the molecules to be reproduced by a Markov model. The essential input for the Markov model is the transition matrix, obtained from a suitable discretization of the configuration space trajectory. The features of the configuration space networks are analyzed in terms of the fractal dimensions. The predictions of the network models for the kinetics are compared to the original MD trajectories. Secs. 5.3 and 5.4 contain the principal results obtained in the present thesis. Chap. 6 summarizes the results presented and briefly discusses potential perspectives for future work.

*Eng ist die Welt, und das Gehirn ist weit.
Leicht beieinander wohnen die Gedanken,
doch hart im Raume stoßen sich die Sachen.*
FRIEDRICH SCHILLER

CHAPTER 2

THERMAL FLUCTUATIONS

In the following chapter, we give a brief overview of the history of the kinetic theory and introduce the fundamental concepts from statistical mechanics, which form the basis of the present thesis. In particular, we present the theory of Brownian motion, the random walk and the phenomenon of anomalous diffusion.

2.1 Very brief history of the kinetic theory

Joseph Fourier found in 1807 that the conduction of heat is proportional to the temperature gradient. Together with the continuity equation he derived the heat equation for a temperature field, $T(x,t)$,

$$\frac{\partial T}{\partial t} = \kappa \frac{\partial^2 T}{\partial x^2}. \qquad (2.1)$$

Fourier also developed a method to solve the heat equation, the Fourier decomposition of periodic functions [55].

Fourier's work was an important contribution to the physics of heat, although, at the time, it was not yet clear what heat actually is. Independently, James Prescott Joule (in 1843) and Robert Mayer (in 1842) identified heat as a form of energy. On the basis of these works Hermann von Helmholtz established the first law of thermodynamics in 1847 [56]. In 1855 the physiologist Adolf Fick found Fourier's heat equation to govern also the dynamics of diffusing concentrations [2].

The second law of thermodynamics was implicitly formulated by Rudolf Clausius in 1850. He also promoted in 1857 the idea of heat emerging from the kinetic energy of small particles, which he called *molecules*, *i.e.* small constituents of matter [57]. Later, in 1865, he introduced the concept of *entropy* to quantify irreversibility [58]. Although the idea of an atomistic structure of matter was suggested earlier – apart from philosophical speculations, Daniel Bernoulli, for example, had had similar speculations in the 18th century – Clausius' work initiated a new branch in physics, *kinetic theory*, primarily applied to the thermodynamics of gases. The appeal of kinetic theory arose from its power to unify different physical fields: kinetic theory raised the hope of thermodynamics being based on mechanics [56].

In the following years several results of thermodynamics could be reproduced with kinetic theory, like the specific heat of one-atomic gases or the ideal gas law. James Clerk Maxwell derived the first statistical law in physics, the velocity distribution of molecules in a gas. One of the great feats of kinetic theory was the H-theorem of Ludwig Boltzmann in 1872 [59]. The atomic picture led him to conclude that the second law of thermodynamics is merely a statistical law. Therefore, a decrease in entropy is unlikely but not *a priori* impossible. Boltzmann's result was received with great skepticism[1]. As Max Planck remarked[2], Boltzmann did not acknowledge the pivotal role of molecular disorder in his proof of the H-theorem. This inattention may have been the reason why Boltzmann did not attempt to prove kinetic theory on the microscopic level.

Boltzmann also considered different averaging procedures in his work and argued that averages over an ensemble and averages over time must coincide for most, but not all, mechanical systems [61]. Boltzmann assumed a more far-reaching version, the ergodic hypothesis, which was later disproved by Johann von

---

[1] The younger Max Planck criticized Boltzmann as Planck did not accept the statistical nature of the second law. Planck doubted the existence of atoms but later changed his mind. Besides Planck's assistant Ernst Zermelo, who criticized Boltzmann on mathematical arguments, the chemist Wilhelm Ostwald and the physicist-philosopher Ernst Mach were the ones who fiercely denied the existence of atoms. Ostwald claimed to have been convinced otherwise by the results of Perrin's work.

[2] „In der Tat fehlte in der Rechnung von Boltzmann die Erwähnung der für die Gültigkeit seines Theorems unentbehrlichen Voraussetzung der molekularen Unordnung."[60]

Neumann and George David Birkhoff, who established the mathematical theory of ergodicity. However, Boltzmann intuitively captured an important property of a large class of physical systems [56].

As a doctoral student at the University of Zürich, Albert Einstein worked on the measurement of molecular weights. In 1905, he published an article about the motion of particles suspended in a resting fluid [3]. Einstein realized that the thermal motion of the molecules in a liquid, as stated by kinetic theory, gives rise to the motion of suspended beads which are large enough to be visible under a microscope. He proposed this effect as an ultimate touchstone of kinetic theory and conjectured it could be identical to Brownian motion, which had been discovered eighty years earlier [1]. At the same time, Marian von Smoluchowski at Lemberg worked on similar questions [62]. The concept of random walks, a term coined by Karl Pearson, was discussed in the journal Nature [63]. In 1908, Paul Langevin developed the idea of stochastic differential equations [64], a seminal approach to Brownian motion and other problems involving randomness.

Einstein's work laid the basis for Jean Perrin's observations of Brownian particles in 1909 [4]. Perrin, an experimentalist from Paris, had developed a new type of microscope that allowed him to measure the Brownian motion with a precision previously unattained. His validation of Einstein's theoretical predictions was among the results which earned him the Nobel Prize in physics in 1926. The discovery of the thermodynamic origin of Brownian motion was the first quantitative observation of microscopic thermal fluctuations.

2.2 Basic concepts

Next, we turn our attention to the basic concept employed in statistical mechanics and diffusion theory.

Let a system consist of N particles with masses m_i. The Cartesian positions of the particles are merged to the $3N$ position vector, $\boldsymbol{r}(t)$. The coordinates are chosen such that the center of mass is always at the origin. The velocities form the $3N$ vector, $\boldsymbol{v}(t)$. The $6N$ dimensional space of the combined position-velocity vector is the phase space, sometimes denoted as Γ and referred to as Γ-space.

At each time, t, the point in the phase space, $(\boldsymbol{r}(t), \boldsymbol{v}(t))$ defines the *state* of the system. Let $\mathbf{M}$ be the diagonal mass matrix with the components $M_{ij} = \delta_{ij} m_j$. The time evolution of the system is given by Newton's second principle

$$\mathbf{M} \frac{\mathrm{d}^2}{\mathrm{d}t^2} \boldsymbol{r} = \boldsymbol{f}, \tag{2.2}$$

or by equivalent schemes such as Lagrange's or Hamilton's formalism. The solution $(\boldsymbol{r}(t), \boldsymbol{v}(t))$ can be seen as a trajectory in the phase space which is parametrized by the time, t. These phase space trajectories do not intersect. The phase space is filled by a vector field which is analogous to the flow field of an incompressible fluid, as seen from the Liouville theorem [65].

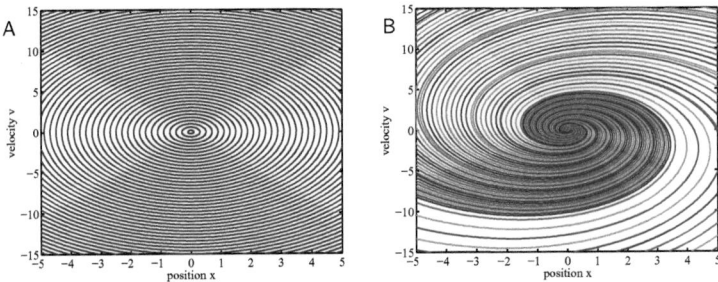

Figure 2.1: **Phase space of the harmonic oscillator.** - A: The phase space trajectories of the harmonic oscillator are ellipsoid and form closed orbits due to the periodicity of the process. B: Phase space trajectories of the harmonic oscillator with friction (underdamped case). All trajectories oscillate but approach the equilibrium $(x, v) = (0, 0)$ for $t \to \infty$. Note that adding friction violates the conservation of energy.

As an example, the one-dimensional harmonic oscillator with angular frequency ω has the equations of motion

$$m\dot{v} = -m\omega^2 r \tag{2.3}$$
$$\dot{r} = v, \tag{2.4}$$

where the dots correspond to time derivatives. These equations define an elliptic field in the phase space, cf. Fig. 2.1. If the state of the system at some point is

2.2 Basic concepts

known, *i.e.*, if the position and velocity are known at some time, the trajectory of the oscillator is defined.

Often, however, it is difficult to describe the definite state of the system, as this would require complete knowledge of all positions and velocities. Rather, the state of the system is described by a probability density, $\rho(\boldsymbol{r}, \boldsymbol{v}, t)$, characterizing the regions of phase space in which the system is likely to be found. The probability density defines an *ensemble*. The members of the ensemble are referred to as *microstates*; the distribution ρ represents a *macrostate* of the system[3]. Statistical mechanics provides the theoretical framework for dealing with phase space probability densities. It bases macroscopic quantities, like temperature and pressure, on the microscopic probability density [65].

For a quantity, $\phi(\boldsymbol{r}, \boldsymbol{v})$, depending on the microstate of the system, the ensemble average is defined as

$$\langle \phi \rangle_{ens} = \int_\Gamma \phi(\boldsymbol{r}, \boldsymbol{v}) \rho_{eq}(\boldsymbol{r}, \boldsymbol{v}) \mathrm{d}\Gamma, \qquad (2.5)$$

where the integration over the Γ-space is denoted as $\mathrm{d}\Gamma = \mathrm{d}^{3N}r \mathrm{d}^{3N}v$.

Imagine a gas of rigid spheres whose only interactions are collisions. Assume the gas is enclosed in a box of finite volume with elastically reflecting walls at constant temperature. The path traveled by an individual particle, its *trajectory*, is a zig-zag line: when it collides with a second particle it changes its velocity and is scattered in another direction until it bumps into the next particle and so on. The microscopic picture shows "that the real particles in nature continually jiggling and bouncing, turning and twisting around one another," as Richard Feynman described it [67]. The permanent agitation brings about temporal variations of observable quantities, the *thermal fluctuations*.

The fluctuation of a quantity, $\phi(\boldsymbol{r}, \boldsymbol{v})$, depending on the positions and velocities, is defined as

$$\langle \Delta \phi^2 \rangle = \left\langle [\phi(\boldsymbol{r}, \boldsymbol{v}) - \langle \phi(\boldsymbol{r}, \boldsymbol{v}) \rangle_{ens}]^2 \right\rangle_{ens}. \qquad (2.6)$$

---

[3] Strictly speaking, the definition of a microstate requires establishing a probability measure in the phase space, Γ. The measure on the phase space emerges from a coarse-graining or a concept of relevance [66].

All microscopic degrees of freedom are fluctuating as a consequence of thermal agitation. A key result from kinetic theory is the *equipartition theorem*. It states that in macroscopic equilibrium every microscopic degree of freedom entering the Hamilton function quadratically, undergoes fluctuations such that the corresponding energy equals $k_B T/2$, where k_B is the Boltzmann constant and T the thermodynamic temperature [65, 68, 69].

Hence, a gas or liquid is never at rest on a molecular level[4]; molecules in a gas or liquid keep themselves permanently agitated. The temperature T is a measure of the energy content per degree of freedom.

Since the average velocity of an arbitrary particle is zero, applying the equipartition theorem, the average kinetic energy per molecule along one direction is

$$\frac{1}{2} m \left\langle \Delta v^2 \right\rangle = \frac{1}{2} k_B T. \qquad (2.7)$$

Note that the particle has this energy in all three spacial dimensions, so the total kinetic energy per particle is three times as much as in Eq. (2.7). In the equipartition theorem the notion of *macroscopic equilibrium* deserves some attention. A thermodynamic equilibrium state has the property of being constant in time. The second law states that thermodynamic processes evolve in time towards an equilibrium state. A system that has reached equilibrium will not leave this state unless an external perturbation occurs. The microscopic picture of kinetic theory assumes molecules that are permanently moving. Only the macroscopic properties do not vary in the equilibrium state.

The one-dimensional distribution of velocities v of the particles in an ideal gas, each of mass m, at temperature T is given by the Maxwell-Boltzmann distribution

$$P_{mb}(v) = \sqrt{\frac{m}{2\pi k_B T}} \exp\left(-\frac{mv^2}{2k_B T}\right). \qquad (2.8)$$

This distribution is stationary under the random scattering of the gas molecules, as can be demonstrated with the Boltzmann equation. The latter involves the assumption of molecular disorder which allows to neglect the correlations of two particles just having collided (molecular chaos). Besides Eq. (2.8), there is no

---

[4] The case of a temperature $\lim T \to 0$ can be treated correctly only in the framework of quantum mechanics – a subject which is not touched in the present thesis.

2.2 Basic concepts

other distribution that is stationary with respect to the Boltzmann equation. Therefore, Eq. (2.8) gives the equilibrium distribution of velocities of an ideal gas, or of any system that can be described by the Boltzmann equations together with the molecular chaos assumption.

Ergodicity

Besides the ensemble average, there is a second way to perform an average: time averaging. For a quantity $\phi(\tau)$ the time average is defined as

$$\langle \phi(\tau) \rangle_\tau = \lim_{T \to \infty} \frac{1}{T} \int_0^T \phi(\tau) \mathrm{d}\tau. \qquad (2.9)$$

A Hamiltonian system with constraints is restricted to a certain subset or surface[5], S, in the Γ-space. The subset, S, reflects the total energy and constraints excluding the "forbidden" points in the phase space. Boltzmann assumed for a Hamiltonian system that the system visits every microstate on the surface S, i.e., every accessible[6] microstate, after sufficient time. That is, every phase trajectory crosses every phase space point which is not in conflict with the constraints or boundary conditions, and all accessible microstates are equally probable. This is the so-called *ergodic hypothesis*. In the strict sense, the hypothesis is wrong. However, it can be proven that all phase space trajectories of a Hamiltonian system, except a set of measure zero, come arbitrarily close to every accessible point in phase space (quasi-ergodic hypothesis) and that the time spent in a region of the surface is proportional to the surface area in the limit of infinitely long trajectories. Thus, for a Hamiltonian system, ensemble averages and time averages yield the same results. In general, systems that exhibit the same behavior in time averages as in ensemble averages are called *ergodic* [65].

For all practical applications, there is a finite observation time, T and the limit procedure in equation Eq. (2.9) cannot be carried out for a real experiment

---

[5] If the only constraint is the total energy, the dynamics in Γ-space are restricted to a $6N - 1$-dimensional subset, *i.e.*, a hypersurface in phase space. However, the term surface is used even if more constraints exist.

[6] A state is called "accessible" if it is not in conflict with the constraints of the system – *e.g.*, constant total energy – and if it is connected, *i.e.*, if there is a phase space path connecting the state with the initial condition which does not violate the constraints.

or for a computer simulation. The following notation will be used to denote finite time averages

$$\langle \phi(\tau) \rangle_{\tau,T} = T^{-1} \int_0^T \phi(\tau) \mathrm{d}\tau. \qquad (2.10)$$

The validity of equating $\langle \cdot \rangle_\tau$ with $\langle \cdot \rangle_{\tau,T}$ holds, of course, only if T is "large enough". In principle T must be larger than every timescale innate to the system or quantity of interest. Whenever the time T is too short, the system appears to be non-ergodic, irrespective of whether the system is ergodic or not.

2.3 Brownian motion

Small, but microscopically observable particles suspended in a liquid undergo a restless, irregular motion, so-called Brownian motion.

In 1908, Langevin developed a scheme to model Brownian motion[7] which turned out to be an approach of general applicability [64, 69, 70]. He formulated the following stochastic differential equation for the one-dimensional movement of a microscopic particle of mass m

$$m \frac{\mathrm{d}^2}{\mathrm{d}t^2} x(t) = -m\gamma \frac{\mathrm{d}}{\mathrm{d}t} x(t) + \xi(t). \qquad (2.11)$$

The first term on the right accounts for the friction with the sourrounding particles arising from the collisions with these particles. The friction constant, γ, determines the strength of the interaction between the Brownian particle and its environment, and is linear in velocity $v = \mathrm{d}x/\mathrm{d}t$ corresponding to Stokes's friction law. The given expression is just an averaged value; deviations from that value are expressed by the random force ξ. This random force keeps the particle moving. If there were only friction, the particle would come to rest. Eq. (2.11) is called the *free* Langevin equation, because there is no force field present. A term representing the force due to a space dependent potential $V(x)$, *e.g.*, a drift or a

---

[7] The term *Brownian motion* is used in different contexts with slightly different meanings. Here, it is used to describe the microscopic motion of suspended particles, *i.e.*, the original phenomenon observed by R. Brown and originally referred to as "Brown'sche Molecularbewegung" by Einstein, and in a broader sense as the process described by the Langevin equation Eq. (2.11).

2.3 Brownian motion

confining harmonic potential, can be added to Eq. (2.11). These terms sum up to the total force that acts on the Brownian particle.

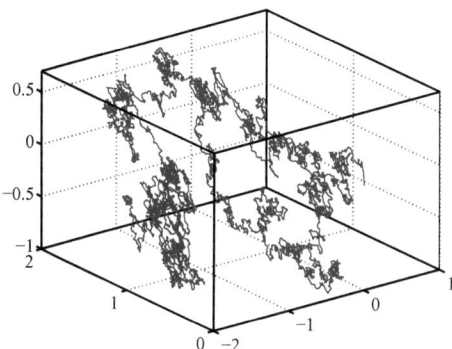

Figure 2.2: **Brownian motion.** - Three-dimensional realization of Brownian motion. The data are obtained from a numerical integration of the Langevin equation, Eq. (2.11), with the Beeman algorithm [71].

The random force $\xi(t)$ – also referred to as *noise* – is assumed to be a stationary random variable. The term *stationary* refers to a quantity or a situation with invariance under time shift. For the random force, $\xi(t) \in \mathbb{R}$, this means: ξ is drawn from a distribution $P(\xi)$ that does not depend on the time t. As a consequence, the process itself is stationary. The distribution $P(\xi)$ is assumed to have a zero mean

$$\langle \xi \rangle = \int \xi P(\xi) \mathrm{d}\xi = 0, \qquad (2.12)$$

that is, there is no drift due to the random force. Furthermore, we assume

$$\langle \xi(t+\tau)\xi(\tau) \rangle = 2B\delta(t). \qquad (2.13)$$

Hence, the variance of $P(\xi)$ is assumed to be $\langle \xi^2 \rangle = 2B$. From Eq. (2.13) it is obvious that, for the Langevin random force ξ, no time correlations exist. The term *white noise* refers to this lack of time correlations in ξ. Note that Eq. (2.13) does not depend on the time τ, due to the time-independence of $P(\xi)$. Commonly, it is assumed that ξ is a Gaussian random variable, an assumption that can be justified by the central limit theorem in many applications.

Eq. (2.11) is the stochastic analog of Newton's equation of motion. The averaging procedure in Eqs. (2.12) and (2.13) is to be understood as performed over the random force ξ itself. As the Langevin equation Eq. (2.11) represents a 'typical' particle in the ensemble, this averaging procedure corresponds to an ensemble average. More precisely, the definition of the random force ξ, which is the only random variable of the process, determines a specific ensemble: the equation of motion, Eq. (2.11) together with the noise, gives rise to a certain population of the phase space.

From Eq. (2.11) the velocity is obtained as

$$v(t) = v_0 e^{-\gamma t} + e^{-\gamma t} \int_0^t e^{\gamma \tau} \frac{\xi(\tau)}{m} d\tau, \qquad (2.14)$$

in which $v_0 = v(0)$ is the initial velocity. After taking the square of Eq. (2.14) and performing the average over the noise, the following expression is obtained

$$\langle v^2(t) \rangle = v_0^2 e^{-2\gamma t} + \frac{B}{\gamma m^2}, \qquad (2.15)$$

where the cross term vanished due to $\langle \xi \rangle = 0$, Eq. (2.12). The velocity reaches a constant value for long times and from the second law it follows that $t \to \infty$ corresponds to the equilibrium, in which the initial condition is entirely forgotten. Therefore, the time $(2\gamma)^{-1}$ can be seen as a relaxation time. In thermal equilibrium, the equipartition theorem can be applied. Under these conditions, the constant B, introduced in Eq. (2.13), can be determined as $B = m\gamma k_B T$. Hence, the correlations of the noise are linear in the temperature and in the friction coefficient. This result is commonly referred to as the *fluctuation-dissipation theorem*.

The friction represents the interaction between the Brownian particle and the environment – but only the dissipative part of it. The thermal agitation also gives rise to the random force, a measure of which is given by the temperature; it keeps the particle moving. Hence, with increasing T, the agitation becomes stronger, which is accounted for by the increased variance of ξ. Therefore, the form of B is a consequence of the conservation of energy. The fluctuation-dissipation theorem assures the conservation of energy: the energy dissipated due to the friction equals the energy transferred to the system by the random force. The fluctuation-dissipation theorem is valid only at equilibrium conditions.

2.3 Brownian motion

In the time interval $[t_1, t_2]$ the Brownian particle travels a certain distance, called the displacement. If the Langevin process is assumed to have no drift, as e.g., in Eq. (2.12), the motion is symmetric. Hence, the mean displacement equals zero.

A quantity of interest is the mean squared displacement (MSD), defined as

$$\langle \Delta x^2(t) \rangle = \langle [x(t+\tau) - x(\tau)]^2 \rangle. \tag{2.16}$$

The MSD can be calculated as a time average or as an ensemble average. The displacement from the initial position x_0 up to time t is $x(t) - x_0 = \Delta x(t) = \int_0^t v(\tau)\mathrm{d}\tau$. Performing the integration in Eq. (2.14), the noise averaged MSD reads [70]

$$\langle \Delta x^2(t) \rangle_\xi = \frac{v_0^2}{\gamma^2}\left(1 - \mathrm{e}^{-\gamma t}\right)^2 + \frac{k_B T}{\gamma^2 m}\left(2\gamma t - 3 + 4\mathrm{e}^{-\gamma t} - \mathrm{e}^{-2\gamma t}\right). \tag{2.17}$$

After averaging over the initial velocity given by the Maxwell-Boltzmann distribution in Eq. (2.8), i.e. $\langle v_0^2 \rangle_{eq} = k_B T/m$, the equilibrium expression is obtained as [69, 70]

$$\langle \Delta x^2(t) \rangle_{eq} = \frac{2k_B T}{m\gamma^2}\left(\gamma t - 1 + \mathrm{e}^{-\gamma t}\right). \tag{2.18}$$

For short t, the MSD is quadratic in t. The ballistic behavior is a consequence of inertial effects: for short times the particle travels in a straight line with nearly constant velocity. For long t, when inertia has been dissipated, the MSD exhibits a linear time dependence.

Langevin equation with harmonic potential

The Langevin equation allows an external force field to be included in Eq. (2.11). In the following, a one-dimensional Langevin equation with harmonic potential is discussed,

$$m\frac{\mathrm{d}^2}{\mathrm{d}t^2}x(t) = -m\gamma\frac{\mathrm{d}}{\mathrm{d}t}x(t) - m\omega^2 x(t) + \xi. \tag{2.19}$$

The Langevin equation with harmonic potential describes an ergodic system, i.e., for a quantity $\phi(x)$ that depends on x, which is itself governed by Eq. (2.19),

$$\langle \phi(x) \rangle_\tau = \left\langle \langle \phi(x) \rangle_\xi \right\rangle_0. \tag{2.20}$$

The average $\langle \cdot \rangle_0$ represents an average over equilibrium initial conditions[8], as is performed to obtain Eq. (2.18). The equilibrium distribution of the velocity is given by the Maxwell-Boltzmann distribution, Eq. (2.8), the coordinate equilibrium distribution obeys Boltzmann statistics,

$$P_{eq}(x) = \frac{\omega\sqrt{m}}{\sqrt{2\pi k_B T}} \exp\left(-\frac{m\omega^2 x^2}{2k_B T}\right). \tag{2.21}$$

Eq. (2.20) is the *ergodic theorem* for the Langevin equation. In what follows, we do not differentiate between time and equilibrium ensemble averages, as is justified due to the ergodicity of Eq. (2.19).

In the case of the harmonic Langevin equation, the equipartition theorem for the kinetic energy is the same as in Eq. (2.7). Additionally, the equipartition theorem can be applied to the vibrational energy,

$$\frac{1}{2}m\omega^2 \langle x^2 \rangle_\tau = \frac{1}{2} k_B T. \tag{2.22}$$

In the expression of the mean squared velocity, Eq. (2.15), derived for the free Langevin equation, the system exhibits a memory of the initial value of the velocity. The Langevin process contains inertial effects due to the present state of the system being influenced by past states: the dynamics are correlated. This phenomenon is quantified by the auto-correlation function (ACF), which can be calculated for various quantities. Note that correlations can arise not only from inertia, but also from ignorance against dynamical details. A more general treatment of projection procedures is given in Sec. 5.1.

In Eq. (2.13) the ACF of the random force is given. Note that we did not specify an initial condition for ξ and Eq. (2.13) is invariant with time shift, *i.e.*, the right side of the equation does not depend on τ. Other common ACFs are the coordinate ACF[9] (CACF), given by

$$C_x(t) = \frac{\langle x(t+\tau)x(\tau) \rangle_\tau}{\langle x^2(\tau) \rangle_\tau}, \tag{2.23}$$

---

[8] Such an average requires a uniquely defined equilibrium state. However, there is no equilibrium distribution of the coordinate in the free, unbounded case in Eq. (2.11). Thus, for quantities depending explicitly on the distribution of initial x values, there is also no equilibrium. Therefore, the ergodic theorem cannot be applied. Furthermore, a time average over any time interval will never converge, as there is no equilibrium to converge to.

[9] As pointed out above, time averaging requires an equilibrium state to converge. That is

2.3 Brownian motion

and the velocity auto-correlation function (VACF)

$$C_v(t) = \frac{\langle v(t+\tau)v(\tau)\rangle_\tau}{\langle v^2(\tau)\rangle_\tau}. \qquad (2.24)$$

The two ACFs are normalized; the denominator is given by the equipartition theorem. The ACFs quantify how strongly the present state is influenced by the history of the system. However, the influence is a statistical correlation which is not to be confused with a causal relation. If the ACFs decay exponentially, the typical time scale is a relaxation time. Finite time averages must exceed the relaxation times significantly for the ergodic theorem to be applicable.

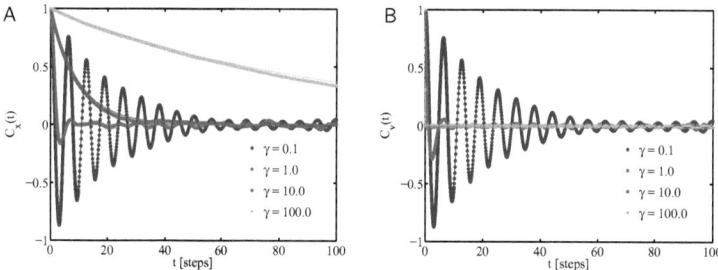

Figure 2.3: (Color online) **Autocorrelation functions of the Langevin process.** - Numerical integration of the Langevin equation with a harmonic potential ($\omega = 1.0$), Eq. (2.19), to calculate the CACF and VACF. The integration is done with the Beeman algorithm [71] for different friction values $\gamma = 0.1, 1.0$ (underdamped) and $\gamma = 10.0, 100.0$ (overdamped). A: CACF, simulation results dotted. The analytical CACF, Eq. (2.25), is given as continuous line. B: VACF, simulation results dotted. The analytical VACF, Eq. (2.26), is given as continuous line.

The CACF and VACF of the Langevin process with harmonic potential read, respectively [72, 73]

$$C_x(t) = \frac{\gamma+\varpi}{2\varpi}e^{-(\gamma-\varpi)t/2} + \frac{-\gamma+\varpi}{2\varpi}e^{-(\gamma+\varpi)t/2} \qquad (2.25)$$

---

why the CACF of the free, unbounded Langevin equation is not defined. There have been cases in which this problem led to confusion.

Note that the MSD does not depend on the initial position and is properly defined for the free, unbounded Langevin equation.

and

$$C_v(t) = \frac{-\gamma + \varpi}{2\varpi} e^{-(\gamma-\varpi)t/2} + \frac{\gamma + \varpi}{2\varpi} e^{-(\gamma+\varpi)t/2}, \qquad (2.26)$$

where the characteristic frequency $\varpi = (\gamma^2 - 4\omega^2)^{1/2}$ is used. The VACF of the free Langevin process can be obtained from Eq. (2.26) in the limit $\omega \to 0$. Note that the CACF is undefined for the free Langevin particle.

There are mainly two different cases to discuss, which are illustrated in Fig. 2.3.

- **Overdamped case** $\gamma > 2\omega$: In this case, ϖ is a real number. Since $\omega > 0$, we have $\varpi < \gamma$. As a consequence all terms in Eqs. (2.25) and (2.26) decay exponentially to zero for large t. The leading term in the CACF is $\exp[-(\gamma - \varpi)t/2]$. In the limit $\gamma \gg 2\omega$, this term will be constantly equal to one, while the second term in Eq. (2.25) is zero then, due to its vanishing coefficient. Note that the VACF does not show such particular behavior. The dominating term in the VACF is $\exp[-(\gamma - \varpi)t/2]$, but its coefficient approaches zero as the friction increases. So for $\omega \to 0$, the VACF is equal to its second term, given by $e^{-\gamma t}$.

- **Underdamped case** $\gamma < 2\omega$: In this case, the frequency ϖ is an imaginary number. This leads to oscillatory behavior of the CACF and the VACF, which will be bounded by the damping factor $\exp(-\gamma t/2)$, giving rise to an exponential decay.

The question arises as to how the CACF is connected to the MSD. Given that the CACF exists, one obtains from the definition of the time-averaged MSD

$$\left\langle \Delta x^2(t) \right\rangle_\tau = 2 \left\langle x^2 \right\rangle_\tau [1 - C_x(t)]. \qquad (2.27)$$

Therefore, the CACF and MSD contain essentially the same information.

2.4 From random walks to diffusion

"Can any of your readers refer me to a work wherein I should find a solution of the following problem, or failing the knowledge of

2.4 From random walks to diffusion

any existing solution provide me with an original one? I should be extremely grateful for aid in the matter."

A man starts in the point O and walks l yards in a straight line; he then turns through any angle whatever and walks another l yards in a second straight line. He repeats this process n times. I require the probability that after these n stretches he is at a distance between r and $r + \delta r$ from his starting point, O.

This question of the British mathematician Karl Pearson is entitled *The problem of the random walk* and was published in the Nature magazine in July 1905 [63].

In the following, Pearson's question is answered for the one-dimensional random walk. If the walker arrives in x at the time t he moves to $x + \Delta x$ or $x - \Delta x$ after a fixed time interval Δt. The probability of moving in the positive or negative direction may be $1/2$ each. Let $W(x,t)$ be the probability of finding the walker in the range $[x, x + \Delta x]$ at time t. This probability is a distribution of the random variable x, while t is just a parameter. $W(x,t)$ is a probability linked to the ensemble average. Thus, the ensemble average of a x-dependent quantity $f(x)$ reads

$$\langle f(x) \rangle_{ens} = \int_{-\infty}^{\infty} f(x) W(x,t) \mathrm{d}x. \tag{2.28}$$

For the probability $W(x,t)$ the following relation holds

$$W(x, t + \Delta t) = \frac{1}{2} W(x - \Delta x, t) + \frac{1}{2} W(x + \Delta x, t). \tag{2.29}$$

Expanding the probability in a Taylor series for small Δt around t and small Δx around x, respectively, yields

$$W(x, t + \Delta t) = W(x,t) + \Delta t \frac{\partial}{\partial t} W(x,t) + \mathcal{O}(|\Delta t|^2) \tag{2.30}$$

and

$$W(x \pm \Delta x, t) = W(x,t) \pm \Delta x \frac{\partial}{\partial x} W(x,t) + \frac{(\Delta x)^2}{2} \frac{\partial^2}{\partial x^2} W(x,t) + \mathcal{O}(|\Delta x|^3). \tag{2.31}$$

Inserting the two Taylor expansions in Eq. (2.29) leads to

$$\frac{\partial}{\partial t} W(x,t) + \mathcal{O}(|\Delta t|) = \frac{(\Delta x)^2}{2\Delta t} \frac{\partial^2}{\partial x^2} W(x,t) + \mathcal{O}(|\Delta x|^3 |\Delta t|^{-1}). \tag{2.32}$$

At the limits $\Delta t \to 0$ and $\Delta x \to 0$, such that $D = (\Delta x)^2/2\Delta t$ has a finite, non-zero value, the diffusion equation follows as

$$\frac{\partial}{\partial t} W(x,t) = D \frac{\partial^2}{\partial x^2} W(x,t). \tag{2.33}$$

The constant D is referred to as *diffusion constant*.

The diffusion equation is not restricted to the particular setup of the one-dimensional random walk. It can also be obtained as a continuum limit of the Langevin process and a large class of random walks with continuous jump lengths and irregular jump frequency, cf. Eq. (D.58) in Appendix D. Eq. (2.33) is valid when no external force fields are present. This field-free case was implicitly assumed when the probability of moving to the positive and negative direction was uniquely set to 1/2 for either direction. The diffusion equation can be extended to processes with external force fields leading to the Fokker-Planck equation [65, 69].

If Eq. (2.33) is applied to all $x \in \mathbb{R}$ (free, unbounded case), an analytical solution can be obtained. With the initial condition $W(x,0) = \delta(x)$, the solution is given as

$$W(x,t) = \frac{1}{\sqrt{4\pi D t}} \exp\left(-\frac{x^2}{4Dt}\right). \tag{2.34}$$

The MSD can be obtained from this probability distribution as

$$\left\langle \Delta x^2(t) \right\rangle_{ens} = \int_{-\infty}^{\infty} x^2 W(x,t) \mathrm{d}x = 2Dt. \tag{2.35}$$

Eq. (2.35) corresponds to the long-time, asymptotic behavior of the Langevin MSD Eq. (2.18). The short-time, ballistic behavior is not reproduced by Eq. (2.33). From the comparison with the large-t behavior of Eq. (2.18), it follows

$$D = \frac{k_B T}{m\gamma}. \tag{2.36}$$

This identity is known as the *Einstein relation*, which relates the macroscopic observable, D, with the microscopic quantity $m\gamma$ [3].

Fourier decomposition allows the diffusion equation to be solved, including various types of boundary conditions. Examples are given in Appendix A.

2.5 Anomalous diffusion

The consideration of the Langevin process revealed three ranges of the MSD. For short times, the inertia of the Langevin particle causes a ballistic behavior with a quadratic time dependence of the MSD. After the friction has dissipated the inertial contribution, the particle's MSD exhibits a linear time dependence. If the accessible volume is finite or, *e.g.*, a harmonic potential is present, the MSD saturates for long times. As the MSD is a continuous function of the time, there are cross over regions between the three 'phases' of the MSD. However, the cross over regions extend over short time ranges.

In contrast to the MSD of a Langevin particle, the MSD obtained in various experiments exhibits a time dependence different from that of all three typical phases. Several experiments from different scientific disciplines reveal a power-law time dependence of the MSD,

$$\left\langle \Delta x^2(t) \right\rangle \sim t^\alpha, \qquad (2.37)$$

with an exponent unequal to one, which cannot be explained as a cross-over effect. α is referred to as MSD exponent. Processes with a power-law MSD, in which $\alpha \neq 1$, are referred to as *anomalous diffusion*. If the exponent is larger than one, the MSD is said to be *superdiffusive*. Likewise, a MSD with an exponent $0 < \alpha < 1$ is called *subdiffusive*. From Eq. (2.27) it can be seen that a power-law behavior applies to both, the MSD and the CACF. Therefore, both the MSD and the CACF serve as indicators of anomalous diffusion.

The main interest of the thesis at hand is subdiffusive processes. In the following, some hallmark experimental results revealing subdiffusive dynamics are presented. The list is by far not exhaustive. Further examples can be found in [8, 74].

- Harvey Scher and Eliott W. Montroll worked in the early 1970s on the charge transport in amorphous thin films as used in photocopiers. The transient photo-current in such media is found to decay as a power-law, indicating persistent time correlations [9].

- The transport of holes through Poly(p-Phenylene Vinylene) LEDs has been observed to be subdiffusive with a subdiffusive exponent $\alpha = 0.45$ [75]. The exponent does not depend on the temperature and the subdiffusion is hence seen as a consequence of structural disorder.

- Financial data, such as the exchange rate between US dollar and Deutsche Mark, can be described in terms of a subdiffusive process [76, 77].

- The diffusion of macromolecules in cells is affected by crowding, *i.e.*, by the presence of other macromolecules in the cytoplasm such that the MSD is subdiffusive [78–82]. The polymeric or actin network in the cell is an obstacle for the diffusing molecules, causing their dynamics in the cytoplasm to be subdiffusive [83, 84].

- The transport of contaminants in a variety of porous and fractured geological media is subdiffusive [85, 86].

- The spines of dendrites act as traps for propagating potentials and lead to subdiffusion [87].

- Translocation of DNA through membrane channels exhibits a subdiffusive behavior [88].

- Anomalous diffusion is seen in the mobility of lipids in phospholipid membranes [89].

The above examples demonstrate the ubiquity of subdiffusive behavior [90].

In Chap. 3, the methods applied to perform MD simulations and to analyze the data generated by MD simulations are explained. In Chap. 4 we present MD simulation results demonstrating the presence of subdiffusion in the internal dynamics of biomolecules. Various models provide subdiffusive dynamics. Those models, which are candidates for the internal, molecular subdiffusion are discussed in Chap. 5.

Now, we turn to the methods which are used to perform simulations of the time evolution of biomolecules. We give an overview of the established algorithms and introduce some of the standart analysis tools. We also discuss some fundamental issues of simulation, in particular the question of convergence and statistical significance.

Hence, our trust in Molecular Dynamics simulation as a tool to study the time evolution of many-body systems is based largely on belief.

D. FRENKEL & B. SMIT

CHAPTER 3

MOLECULAR DYNAMICS SIMULATIONS OF BIOMOLECULES

In this chapter, we present the methods which are commonly used to simulate the time evolution of biomolecular systems with computers. Furthermore, some of the analysis tools established in the field are introduced.

The dynamics of molecules can be described by the quantum mechanical equations of motion, *e.g.*, the Heisenberg or the Schrödinger equation. However, due to environmental noise, quantum mechanical effects can be approximated in many cases by their classical counterparts [91]. Still, for molecules containing many atoms, the classical equations of motion are too difficult to be treated by analytical means, due to their overall complexity [92, 93]. Only computer simulations enable us to exploit the dynamical information enclosed in the equations of motion.

There are further benefits from computer simulations. Experimental techniques cannot resolve the dynamics in all details and it is often complicated to manipulate the systems as one wishes. Computer simulations give access to the full atomic details of a molecule. They allow the parameters of the models to be manipulated, and to test models in regions that are inaccessible in experiments.

The concept of a molecular dynamics (MD) simulation is as follows. A number of molecules, *e.g.*, a peptide and surrounding water molecules, is given in an initial position. The interaction between the atoms form the so-called *force field*. Essentially, the force field and other parameters, like boundary conditions or coupling to a heat bath, build the *physical model* of the system. The forces are represented by the vector $\boldsymbol{f}$. If the system consists of N point-like particles, then $\boldsymbol{f}$ has $3N$ components. In the present thesis, the point-like particles in the MD simulation are the atoms. The force vector is derived from the underlying potential,

$$\boldsymbol{f} = -\frac{\partial V(\boldsymbol{r})}{\partial \boldsymbol{r}}, \tag{3.1}$$

where the $\boldsymbol{r}$ is the configuration vector which contains the $3N$ atomic position coordinates, *e.g.* the Cartesian coordinates of all particles. The equations of motion corresponding to the physical model, *i.e.*

$$\mathbf{M}\frac{\mathrm{d}^2 \boldsymbol{r}}{\mathrm{d}t^2} = \boldsymbol{f}, \tag{3.2}$$

where the mass $3N \times 3N$-matrix $\mathbf{M}$ containing the masses of all atoms on the diagonal is used, are integrated numerically. The integration is performed for a certain time. The data produced in this initial part of the simulation cannot be used to analyze the system, as the system needs time to equilibrate. The estimation of the time needed for equilibration is a non-trivial problem [94]. After this equilibration period, the data collected can be used for the analysis. Thus, MD simulations generate a time series of coordinates of the molecules in the system, the so-called *trajectory*. With the initial preparation, the equilibration, and the recording of data, MD simulations are similar to real experiments. Therefore, they are sometimes referred to as *computer experiments*.

In MD simulations, the molecules are assumed to be formed of atoms[1] which exhibit classical dynamics, *i.e.*, the time evolution can be characterized by Newton's second principle or equivalent schemes (Lagrange, Hamilton). The intra- and intermolecular forces are derived from empirical potentials. Non-conservative forces, *e.g.* friction, are absent on the molecular level[2]. The classical mechanics

---

[1] The atoms in MD simulation are treated as point particles. In some sense this is a form of the Born-Oppenheimer approximation [95], in which the electrons are assumed to follow instantaneously the motions of the nuclei. The nuclei are three orders of magnitude heavier than the electrons, so electronic motion can be neglected on the time scale of nuclear motion. Due to the fast electron dynamics, it is a valid approach to separate electron and nuclear dynamics.

[2] The dissipation of energy by friction can be seen as the transition of mechanical energy to

description is in many cases a very good approximation. As long as there are no covalent bonds being broken or formed, and as long as the highest frequencies, ν, in the system are such that $h\nu \ll k_B T$ – h being Planck's constant – there is no need to employ quantum mechanical descriptions. To overcome the problems due to the highest frequencies, corrections in the specific heat and the total internal energy can be included. An alternative is to fix the lengths of covalent bonds. Fixing the bond lengths has the benefit to be more accurate and to allow larger time steps at the same time [96, 97] (see Appendix B).

In order to perform an MD simulation some input is required.

- The initial positions of the atoms must fit the situation of interest. The topology, *i.e.*, the pairs of atoms that have a covalent bond, must be specified. For large biomolecules with secondary structure the conformation is obtained from experiments, *e.g.*, from X-ray crystallography.

- The initial velocities can be provided as input data. Alternatively, random velocities can be generated from the Maxwell-Boltzmann distribution.

- The force field contains the covalent bonds specified by the topology file and other pairwise, non-bonded interactions like van der Waals and electrostatic forces.

- Boundary conditions determine how to deal with the finite size of the simulation box in which the molecules are placed. Additionally, the coupling to a heat bath or a constant pressure is often included.

In the thesis at hand the GROMACS software package is used [97, 98] with the Gromos96 force field [99].

---

thermal energy, *i.e.*, to a disordered, kinetic energy. For macroscopic objects, the mechanical energy of the macroscopic degrees of freedom is transformed by friction into kinetic energy of microscopic degrees of freedom. In the framework of kinetic theory, the molecular level is the most fundamental one. Therefore, mechanical energy on the atomic level cannot be transformed to "more microscopic" degrees of freedom. Hence, the mechanics of atoms evolves without friction forces. However, sometimes friction forces are introduced to model the interaction with a heat bath, see the paragraph *thermostats* in Sec. 3.2

3.1 Numerical integration

In MD simulations, the Verlet algorithm is a common choice for the integration of the equations of motion [18, 100]. The Taylor expansion is the first step in deriving the algorithm. Let a system consist of N particles. The $3N$-dimensional position vector is $r(t)$, and the forces on the particles are given by the $3N$ dimensional vector $f(t)$. Using Eq. (3.2), the Taylor expansion is

$$r(t + \Delta t) = r(t) + \Delta t v(t) + \frac{\Delta t^2}{2}\mathbf{M}^{-1}f(t) + \frac{\Delta t^3}{6}\frac{\partial^3 r(t)}{\partial t^3} + \mathcal{O}(|\Delta t^4|), \quad (3.3)$$

where $v(t)$ is the velocity. Likewise, for an earlier time it is

$$r(t - \Delta t) = r(t) - \Delta t v(t) + \frac{\Delta t^2}{2}\mathbf{M}^{-1}f(t) - \frac{\Delta t^3}{6}\frac{\partial^3 r(t)}{\partial t^3} + \mathcal{O}(|\Delta t^4|). \quad (3.4)$$

The sum of the Eqs. (3.3) and (3.4) is

$$r_{n+1} \approx 2r_n - r_{n-1} + \mathbf{M}^{-1}f_n \Delta t^2, \quad (3.5)$$

where the subscript indices replace the time dependence with the similar notation for all time dependent quantities, $e.g.$, $r(t) = r(n\Delta t) = r_n$. Eq. (3.5) is the Verlet algorithm, with the following properties [18].

- The algorithm is an approximation in the fourth order in Δt.
- It is strictly time symmetric and reversible.
- It does not make explicit use of the velocity, nor does it provide the velocity.
- It has a moderate energy conservation on time scales of a few high-frequency bond vibrations.
- It has a low drift in the total energy on long time scales, $i.e.$, the total energy is well conserved on time scales well above the fastest bond vibrations.

In principal, the algorithm is designed to integrate the equations of motion. It furnishes a trajectory that is a numerical solution of the differential equation, Eq. (2.2), characterizing the system. As Eq. (3.5) is an approximation, there is

3.1 Numerical integration

a difference between the solution obtained from Eq. (3.5) and the exact solution. This error can be decreased by decreasing Δt. The error that follows from Eq. (3.5) is only one source of the deviation from the "true" dynamics. A system with many degrees of freedom and nonlinear interactions, as most MD simulations in practice are, is extremely sensitive to the initial conditions. This sensitivity lies at the heart of chaotic dynamics.

With these problems in mind, the question arises as to the usefulness of MD simulations. First, the aim of MD simulations is usually not to predict precisely how the system evolves starting from a certain initial condition. Instead, statistical results are expected from useful MD simulations. Still, we need to prove that the data generated by MD simulations share the statistics of the underlying equations of motion. There is some evidence that the MD trajectories stay close to some "real" trajectory on sufficiently long time scales; the MD trajectories are contained in a *shadow orbit* around the "true" trajectory [101, 102]. The statement of Frenkel and Smit in the head of this chapter refers to this problem, "Despite this reassuring evidence [...], it should be emphasized that it is just evidence and not proof," [18].

The approximation in Eq. (3.5) does not break the time symmetry, but is rather a reversible approach. Hence, the Verlet algorithm cannot be used when velocity-dependent friction – which would break the time symmetry – is relevant. Conceptually, it is possible to calculate forward as well as backward in time. In practice, this is not feasible as numerical errors will quickly sum up and shift the time evolution to a different orbit.

The Verlet algorithm does not explicitly calculate the velocities. If needed, the velocity can be obtained by

$$\boldsymbol{v}_n = \frac{\boldsymbol{r}_{n+1} - \boldsymbol{r}_{n-1}}{2\Delta t} + \mathcal{O}(|\Delta t^2|). \tag{3.6}$$

This equation is a second order approximation in Δt of the velocity.

Time symmetry is a prerequisite for energy conservation. In classical mechanical systems without friction the total energy is conserved. The approximation of the Verlet algorithm violates energy conservation to some extent. This leads to a moderate energy fluctuation on time scales, which are short and cover only a few integration steps of length Δt. In contrast, there is only a low energy drift on long time scales, a manifest advantage of the Verlet algorithm.

The Leapfrog algorithm

An algorithm equivalent to the Verlet scheme is the *Leap Frog* algorithm which defines the half integer velocity as

$$v_{n+1/2} = \frac{r_{n+1} - r_n}{\Delta t}. \tag{3.7}$$

Hence, the positions are

$$r_{n+1} = r_n + \Delta t v_{n+1/2}. \tag{3.8}$$

Eq. (3.5) reads with the velocities

$$v_{n+1/2} = v_{n-1/2} + \Delta t \mathbf{M}^{-1} f_n. \tag{3.9}$$

If the velocities are required at the same time as the positions, they can be calculated as

$$v_n = \frac{v_{n+1/2} + v_{n-1/2}}{2} = v_{n-1/2} + \frac{\Delta t}{2} \mathbf{M}^{-1} f_n. \tag{3.10}$$

Iterating Eqs. (3.8) and (3.9) is equivalent to the Verlet algorithm.

3.2 Force field

The central input of the model used in MD simulation is the force field. It consists of two different components: (i) the analytical form of the forces, and (ii) the parameters of the forces.

The force field characterizes the physical model which is exploited by the simulation. In the following, we refer to the Gromos96 force field [99], which is the force field used in the simulations discussed in the present thesis. The atoms are represented as charged point masses. The covalent bonds are listed and define the molecules. During the simulation, covalent bonds cannot be formed nor broken.

The Gromos96 force field usually contains three types of interactions.

- Covalent bonds: The forces that emerge from the covalent bonds act between the atoms listed as bonded. The *bonded forces* include bond stretching, angle bending, proper dihedral bending and improper dihedral bending.

3.2 Force field

- Non-bonded interactions: Coulomb attraction and van der Waals forces lead to interactions between arbitrary pairs of atoms, usually applied only to non-bonded pairs. The Coulomb and van der Waals forces are centrosymmetric and pair-additive. In practice, only interactions within a certain radius are taken into account. The forces are derived from the underlying potential.

- Sometimes, additional constraints, such as a fixed end of a polypeptide chain or fixed bond lengths, are included, e.g., to mimic an experimental setup or to increase the time step.

The covalent bond has four terms that contribute to the potential energy. The corresponding coordinates are depicted schematically in Fig. 3.1.

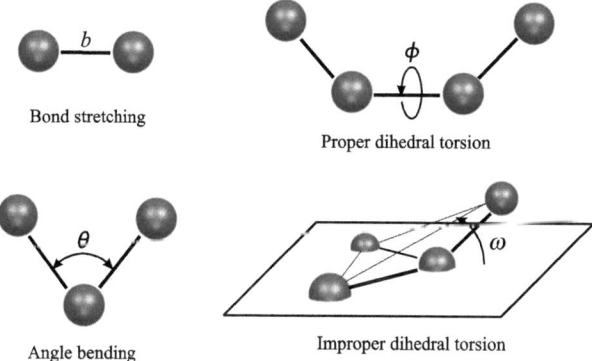

Figure 3.1: **Coordinates used in the force field** - Schematic illustration of the coordinates which are used in the bonded energy terms in equations Eqs. (3.11) to (3.14). Figure from L. Meinhold [103].

- If two bonded atoms, i and j, are separated by a distance b_{ij}, the bond potential energy is

$$V_b = \frac{1}{2}\kappa_b(b_{ij} - b_0)^2, \tag{3.11}$$

which accounts for a stretching of the bond length from the equilibrium value, b_0. The stiffness of the bond is given by the force constant, κ_b.

- The deviation from the equilibrium angle, θ_0, between two neighboring covalent bonds, (i,j) and (j,k), leads to a potential energy of

$$V_\theta = \frac{1}{2}\kappa_\theta(\theta_{ijk} - \theta_0)^2, \qquad (3.12)$$

where θ_{ijk} is the angle between the two bonds. In analogy to Eq. (3.11), κ_θ is the force constant of the harmonic potential (V_θ is harmonic in the angle, not in Cartesian coordinates). The potential V_θ describes a three body interaction.

- The proper dihedral angle potential depends on the position of the four atoms i, j, k, and l, the bonds of which form a chain [see Fig. 3.1]. The angle ϕ_{ijkl} between the plane defined by the positions of i, j, k and the bond k, l, has an equilibrium value ϕ_0. The potential energy due to the torsion of ϕ_{ijkl} is

$$V_\phi = \kappa_\phi[1 + \cos(n\phi_{ijkl} - \phi_0)], \qquad (3.13)$$

a four body interaction.

- The angle between the bond (j,l) and the plane defined by the bonds (i,j) and (j,k) is referred to as improper dihedral, ω_{ijkl}, cf. Fig. 3.1. The torsion of ω_{ijkl} makes the following contribution to the total potential energy.

$$V_\omega = \frac{1}{2}\kappa_\omega(\omega_{ijkl} - \omega_0)^2. \qquad (3.14)$$

The long-range Coulomb force between all pairs of atoms is part of the non-bonded interactions. Its potential has the analytical form

$$V_{es} = \sum_{i,j<i} \frac{q_i q_j}{r_{ij}}, \qquad (3.15)$$

where q_i are the electrical charges and r_{ij} is the distance between atoms i and j, $r_{ij} = |\boldsymbol{r}^i - \boldsymbol{r}^j|$. The second non-bonded interaction is the van der Waals force. Its potential is of Lennard-Jones type, i.e.,

$$V_{vdW} = \sum_{i,j<i} 4\epsilon_{ij}\left[\left(\frac{\sigma_{ij}}{r_{ij}}\right)^{12} - \left(\frac{\sigma_{ij}}{r_{ij}}\right)^6\right], \qquad (3.16)$$

with the depth of the Lennard-Jones potential, ϵ_{ij}, and the collision parameter, σ_{ij}.

The calculation of the non-bonded interactions is the most costly part of the numerical integration. Therefore, the potentials are modified such that they are zero beyond a certain cut-off value. This can be done by a shift function which preserves the force function to be continuous [97]. In order to eliminate unphysical boundary effects, periodic boundary conditions are imposed. The system is contained in a space filling box. In the simulations used in the present thesis, the box is a rhombic dodecahedron or a truncated octahedron. The box is surrounded by translated copies of itself, so-called *images*. As a consequence, boundary effects are avoided. Instead, an unnatural periodicity crops up. The non-bonded, short-range interactions are limited in GROMACS to atoms in the nearest images (minimum image convention). The long-range Coulomb force is treated by methods that are based on the idea of Ewald summation [104], which we now discuss in more detail.

Particle Mesh Ewald Method

In 1921, Paul Peter Ewald, physicist at Munich, developed a scheme to treat the periodic Coulomb interactions in a crystal [104]. In solid state physics it is a common idea to build a crystal from a unit cell and identical images of the unit cell as illustrated in Fig. 3.2 A. The periodic image cells are identified by the vector $\mathbf{n}$. Assume there are N point charges q_i in the unit cell at the positions $\mathbf{r}_i$. The total electrostatic energy including the contributions from the images can be written as

$$V_{es} = \frac{1}{2} \sum_{i=1}^{N} q_i \phi(\mathbf{r}_i), \qquad (3.17)$$

where $\phi(\mathbf{r}_i)$ denotes the potential at the position of q_i, $\mathbf{r}_i$. The potential is given as

$$\phi(\mathbf{r}) = {\sum_{\mathbf{n},j}}' \frac{1}{4\pi\epsilon} \frac{q_j}{|\mathbf{r}_i - \mathbf{r}_j + \mathbf{n}L|}, \qquad (3.18)$$

where the prime denotes that the sum goes over all image cells $\mathbf{n}$, and j over all particles, save the combination $\mathbf{n} = 0$ and $i = j$, as this would correspond to a self interaction.

The sum on the right side of Eq. (3.18) does not generally converge. If it does, it converges slowly. The *Ewald summation* overcomes the slow convergence of the electrostatic potential. The idea is to split Eq. (3.18) into two parts: a quickly converging short-range term and a long-range contribution which can be treated efficiently after it has been transformed into the Fourier k-space.

The separation of the two contributions is achieved by an additional potential that screens each of the point charges q_i with a Gaussian distribution of charge density of opposite sign. The sum of the point charges and the Gaussians is the short range contribution. The virtual charge distribution must be canceled out by a mirror distribution, *i.e.*, the same distribution with opposite sign, see Fig. 3.2 B. This can be handled more efficiently in Fourier space. A third term must be introduced to compensate the self interaction which is due to the energy between the point charge q_i and the mirror of its screening Gaussian.

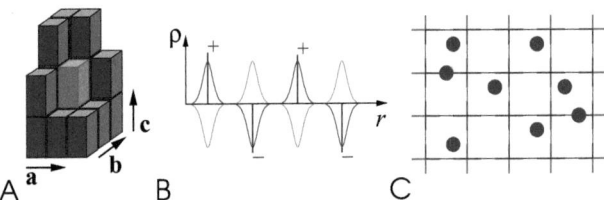

Figure 3.2: *(Color online)* **Calculation of electrostatic interactions in a crystal using the Ewald summation.** - *A:* An ideal crystal is obtained by translations {a,b,c} of a unit cell (red). *B:* In the Ewald summation, each point charge (blue) is surrounded by a neutralizing Gaussian charge distribution (cyan), leading to rapidly convergent series in real space; the corresponding canceling distribution (magenta) is calculated in reciprocal space. *C:* The irregularly distributed charges (blue) are interpolated onto the vertices (green) of a regular grid. Figure and caption from L. Meinhold [103].

Let us start with the mirror charge density given as

$$\rho_m(\boldsymbol{r}) = \sum_j \sum_{\mathbf{n}} q_j \left(\frac{a}{\pi}\right)^{\frac{3}{2}} \exp(-a|\boldsymbol{r}-\boldsymbol{r}_j+\mathbf{n}L|^2). \qquad (3.19)$$

The constant value a specifies the width of the Gaussian. According to the Poisson equation, the charge density gives rise to a potential

$$-\Delta \phi_m(\boldsymbol{r}) = \frac{\rho_m(\boldsymbol{r})}{\epsilon}, \qquad (3.20)$$

3.2 Force field

where Δ is the Laplace operator. The potential corresponding to a point charge, q at the origin is

$$\phi_{pc}(\boldsymbol{r}) = \frac{q}{|\boldsymbol{r}|}. \tag{3.21}$$

It is useful to transform the Poisson equation and the charge density to the Fourier space with the transformation

$$\rho_m(\boldsymbol{r}) = \frac{1}{V} \sum_k \rho_m(\boldsymbol{k}) e^{i\boldsymbol{rk}}, \tag{3.22}$$

where $\boldsymbol{k} = 2\pi \boldsymbol{l}/L$ is the reciprocal cell vector, $\boldsymbol{l} = (l_x, l_y, l_z)$ being the lattice vector in Fourier space. The unit cell is assumed to be cubic with side length L and volume V. The Poisson equation reads in the Fourier representation

$$k^2 \phi_m(\boldsymbol{k}) = \frac{\rho_m(\boldsymbol{k})}{\epsilon}, \tag{3.23}$$

in which $|\boldsymbol{k}|^2 = k^2$. The Fourier transform of the charge density is

$$\rho_m(\boldsymbol{k}) = \frac{1}{V} \sum_j q_j e^{-i\boldsymbol{kr}_j} e^{-k^2/4a}. \tag{3.24}$$

Combining Eqs. (3.23) and (3.24), the potential in Fourier representation is derived for all $\boldsymbol{k} \neq 0$. After inverse transformation, the potential reads

$$\phi_m(\boldsymbol{r}) = \frac{1}{V\epsilon} \sum_{\boldsymbol{k}\neq 0, j} \frac{q_j}{k^2} e^{i\boldsymbol{k}(\boldsymbol{r}-\boldsymbol{r}_j)} e^{-k^2/4a}. \tag{3.25}$$

With the definition $\rho(\boldsymbol{k}) = \sum_j q_j e^{-i\boldsymbol{kr}_j}$, the long range contribution to the total electrostatic energy is expressed as

$$V_{lr} = \frac{1}{2V\epsilon} \sum_{\boldsymbol{k}\neq 0} \frac{1}{k^2} |\rho(\boldsymbol{k})|^2 e^{-k^2/4a}. \tag{3.26}$$

The self-interaction energy, *i.e.*, the energy of q_i due to the mirror potential ϕ_m, is calculated as follows. Both the charge q_i and the Gaussian are shifted to the origin by setting $\boldsymbol{r}_i = 0$. The symmetry allows spherical coordinates to be used. The Gaussian leads to a potential derived from the Poisson equation in spherical coordinates as

$$\phi_{gauss}(\boldsymbol{r}) = q_i \text{erf}(\sqrt{a}r), \tag{3.27}$$

where the error function erf($\cdot$) is used. The self-interaction potential at $\boldsymbol{r}_i$ is given by $\phi_{si}(\boldsymbol{r}_i) = \phi_{gauss}(0) = 2q_i\sqrt{a/\pi}$. The corresponding contribution to the electrostatic energy is

$$V_{si} = -\frac{1}{2}\sum_i q_i\phi_{si}(\boldsymbol{r}_i) = -\sqrt{\frac{a}{\pi}}\sum_i q_i^2. \qquad (3.28)$$

The short range interaction potential is the sum of the potential of the point charge, q_i, and the screening Gaussians,

$$\phi_{sr}(r) = \frac{q_i}{r}\left[1 - \text{erf}(\sqrt{a}r)\right], \qquad (3.29)$$

where r is the radius in spherical coordinates. The energy V_{sr} follows from Eq. (3.29) substituting Eq. (3.17).

Thus, the electrostatic energy can be rewritten as

$$V_{es} = V_{lr} + V_{sr} + V_{si}. \qquad (3.30)$$

Instead of evaluating Eq. (3.18), the numerical implementation of Eq. (3.30) is favorable, as the individual terms of the latter converge much quicker. The algorithm contains a numerical inverse Fourier transform of the sum over the wavevectors $\boldsymbol{k}$ in V_{lr}. This sum scales with the particle number N as $\mathcal{O}(N^2)$. Solving the Poisson equation with a charge distribution on a grid is much faster than the general case. The employment of a grid can be used to speed up the simulation to a numerical efficiency which scales ad $N\log N$ [105, 106]. The charge distribution is interpolated with cardinal B-splines – piecewise polynomial functions which are unequal zero on a small interval (*i.e.* with compact support) –, which allows the Fast Fourier Transform (FFT) to be employed [107] in the evaluation of V_{lr}, see Fig. 3.2 C. The method is referred to as *smooth particle mesh Ewald method* [18, 106].

Water models

In vivo, biomolecules are always found in aqueous solution. The presence of the surrounding water is crucial for proteins to function [42, 43] and it affects the internal dynamics [108]. Water itself is an intrinsically complex substance and its

3.2 Force field

physical properties are an active field of research [109–111]. In the simulations used in the thesis at hand, water is modeled in different ways.

Most of the simulations are performed with an explicit water model, *i.e.*, in addition to the biomolecule of interest the simulation box is filled with MD water molecules using the extended simple point charge (eSPC) water model [112]. It describes water as a tetrahedrally shaped molecule with an OH distance of 0.1 nm. At the positions of the oxygen and hydrogen, point charges of $-0.8467\,e$ and $+0.4238\,e$ are placed, respectively. The effective pair potential between two water molecules is given by a potential of Lennard-Jones type acting on the oxygen positions. The radial distribution that emerges from the eSPC model reproduces the first two peaks of the radial distribution function of oxygen-oxygen distances in real water, the diffusion constant, and density at 300 K. The dipole moment of 2.35 D is higher than the experimental value of 1.85 D [112].

An alternative is to use implicit water models, in which the dynamics of the water is not modeled in atomic detail. One common approach is the generalized Born/surface area (GB/SA) method [113, 114]. To obtain the correct solvation free energy,

$$F_{sol} = F_{cav} + F_{vdW} + F_{pol}, \qquad (3.31)$$

three contributing terms are considered: a solvent-solvent cavity term, F_{cav}, a solvent-solute van der Waals term, F_{vdW}, and a solvent-solute polarization term, F_{pol}. The first two terms are linear in the accessible surface area. The polarization term is obtained from a combination of Coulomb's law with the Born equation, yielding the so-called generalized Born equation. The GB/SA method does not take into account momentum transfer on the solute or friction effects [114]. The model has been compared to explicit water simulations and found to reproduce the dynamics in the case of small peptides [115, 116]. The effect of the solvent can also be included by a Langevin formulation adding a random force and a friction term, cf. Eq. (2.11).

Isokinetic thermostat

The ensemble corresponding to the equations of motion, Eqs. (3.2) and (B.15), has a fixed volume, V, a fixed number of particles, N, and a fixed total energy E.

Experimentally such an ensemble is difficult to establish. Therefore, it is useful to modify the equations of motion and to simulate other ensembles. The isokinetic thermostat [117, 118] allows the $(N, V, \mathcal{T})$ ensemble to be simulated, where $\mathcal{T}$ is the temperature. Let us start with the modified equations of motion

$$\mathbf{M}\frac{d^2\boldsymbol{r}}{dt^2} = \boldsymbol{f} - \beta\boldsymbol{v}, \qquad (3.32)$$

in which β is a constant. The β-term allows the temperature to be kept constant. To see this, we calculate the kinetic energy as

$$3(N-1)k_B\mathcal{T} = \boldsymbol{v}^T\mathbf{M}\boldsymbol{v}. \qquad (3.33)$$

Since the center of mass motion was subtracted, there are $3(N-1)$ degrees of freedom. If the temperature is required to be constant, i.e., $d\mathcal{T}/dt = 0$, and using Eq. (3.32) we have

$$0 = \boldsymbol{v}^T\mathbf{M}\frac{d^2\boldsymbol{r}}{dt^2} = \boldsymbol{v}^T(\boldsymbol{f} - \beta\boldsymbol{v}). \qquad (3.34)$$

From that it follows

$$\beta = \frac{\boldsymbol{v}^T\boldsymbol{f}}{\boldsymbol{v}^T\boldsymbol{v}}. \qquad (3.35)$$

Inserting Eq. (3.32) into the Leap Frog update for the velocity, Eq. (3.9), leads to

$$\boldsymbol{v}_{n+1/2} = \boldsymbol{v}_{n-1/2} + \Delta t \mathbf{M}^{-1}(\boldsymbol{f}_n - \beta\boldsymbol{v}_n). \qquad (3.36)$$

Iterating Eqs. (3.8), (3.10), and (3.36) allows the equations of motion to be integrated at constant temperature. The above approach to model the temperature coupling of the system is designated *isokinetic thermostat*.

3.3 The energy landscape

In Sec. 2.2, the idea of ensembles is introduced. Of particular interest are probability densities that correspond to equilibrium macrostates, i.e., macrostates that do not explicitly depend on time[3],

$$\frac{\partial}{\partial t}\rho_{eq}(\boldsymbol{r}, \boldsymbol{v}, t) = 0 \qquad (3.37)$$

---

[3] For a Hamiltonian system, the total time derivative of the density ρ equals zero. This can be proven from the conservation of probability which leads to a continuity equation for ρ [65].

3.3 The energy landscape

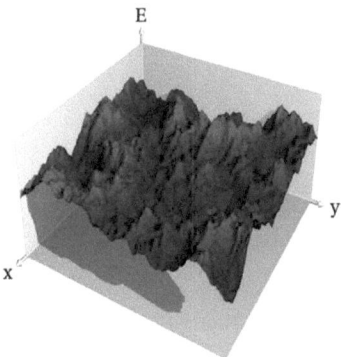

Figure 3.3: **Energy landscape** - Schematic illustration of a rugged energy landscape. The conformation coordinate corresponds to the two-dimensional x-y plane. The potential energy, E, is given on the vertical axis. Figure from Nicolas Calimet.

where ρ_{eq} is the equilibrium phase space density. Hence, for the equilibrium density we skip the time argument, i.e., $\rho_{eq}(\bm{r}, \bm{v})$. Note that from Eq. (3.37) it does not follow that $\frac{\mathrm{d}}{\mathrm{d}t}\rho_{eq}(\bm{r}(t), \bm{v}(t)) = 0$. Nonetheless, averages with respect to the ensemble do not depend on the time, if the ensemble obeys Eq. (3.37). For a molecular system in equilibrium with a constant temperature the distribution of velocities is given by the Maxwell-Boltzmann distribution, $P_{mb}(v)$, Eq. (2.8), irrespective of the potential energy $V(\bm{r})$. Thus, the configuration $\bm{r}$ can be considered as statistically independent from the velocity $\bm{v}$. The statistical independence is expressed by

$$\rho(\bm{r}, \bm{v}) = \rho_r(\bm{r}) P_{mb}(\bm{v}). \tag{3.38}$$

Then, the statistical analysis can be limited to the dynamics in the configuration part of the phase space, the so-called *configuration space*. The potential energy, $V(\bm{r})$ depends on the configuration of the system, the latter represented by $\bm{r}(t)$. Formally, $V(\bm{r})$ is a function that maps the configuration space to real numbers. It can be seen as a hypersurface over the configuration space. This hypersurface is rugged, i.e., it has many local minima separated by barriers, 'peaks' and 'ridges', corresponding to extremely high potential energies, 'valleys' and 'wells' where the energy is low, see Fig. 3.3. These analogies are the reason for referring to the hypersurface as *energy landscape*.

M. Goldstein proposed the idea to describe the dynamics of system in terms of a trajectory on the energy landscape in the context of glass forming liquids [25]. Due to the thermally induced decoupling of position and velocity, the dynamics emerges from the properties of the energy landscape. Since the glass-like aspects of protein dynamics were discovered, the concept of the energy landscape has also been applied to biomolecules [20–22, 26, 119].

3.4 Normal Mode Analysis (NMA)

Let a system of N particles be governed by the equations of motion, Eq. (2.2), and the force be conservative, according to Eq. (3.1). The $3N$ positions are denoted by the $3N$ dimensional vector $\boldsymbol{r}$. Often it is convenient to introduce *mass-weighted coordinates*, i.e.,

$$\boldsymbol{x} = \mathbf{M}^{\frac{1}{2}} \boldsymbol{r}, \tag{3.39}$$

where $\mathbf{M}$ is the diagonal matrix with the components $M_{ij} = \delta_{ij} m_i$ (m_i is the mass of particle i). The equations of motion in the mass-weighted coordinates read

$$\frac{\mathrm{d}^2}{\mathrm{d}t^2} \boldsymbol{x} = \boldsymbol{f}. \tag{3.40}$$

Now assume the potentials of interaction to be purely harmonic, i.e.,

$$V_{harm}(\boldsymbol{x}) = \frac{1}{2}(\boldsymbol{x} - \boldsymbol{x}_0)^T \mathbf{H} (\boldsymbol{x} - \boldsymbol{x}_0), \tag{3.41}$$

where $\mathbf{H}$ is a symmetric, real matrix. The position vector, $\boldsymbol{x}_0$, corresponds to the local minimum of the potential. In the following, it will be assumed that $\boldsymbol{x}_0 = 0$. The coordinates can always be modified such that the minimum is at the origin, $\tilde{\boldsymbol{x}} = \boldsymbol{x} - \boldsymbol{x}_0$. The forces can now be calculated as

$$\boldsymbol{f} = -\frac{\partial}{\partial \boldsymbol{x}} V_{harm}(\boldsymbol{x}) = -\mathbf{H}\boldsymbol{x}. \tag{3.42}$$

Since $\mathbf{H}$ is a symmetric matrix, it can be diagonalized, i.e., there is an orthogonal matrix, $\mathbf{U}$, such that

$$\mathbf{U}^T \mathbf{H} \mathbf{U} = \mathbf{\Theta}, \tag{3.43}$$

3.4 Normal Mode Analysis (NMA)

where Θ is a diagonal matrix with the eigenvalues, θ_i, as non-zero components. The orthogonal matrix $\mathbf{U}$ corresponds to a transformation of coordinates. In the transformed coordinates $\boldsymbol{g} = \mathbf{U}^T \boldsymbol{x}$ the equations of motion read

$$\frac{d^2}{dt^2}\boldsymbol{g} = -\Theta \boldsymbol{g}. \qquad (3.44)$$

As Θ is diagonal, the $3N$ equations of motion are completely decoupled when expressed in the coordinates, $\boldsymbol{g}$. Thus, Eq. (3.44) can be solved for each component $g_i(t)$ separately. The solutions of Eq. (3.44), the $g_i(t)$, are referred to as *normal modes*. Once the dynamics is given in normal modes, it can be inversely transformed to ordinary coordinates $\boldsymbol{r}$. Hence, the dynamics can be understood as a linear combination of the contributions of the individual normal modes. The method to change the coordinates of a harmonic potential and using the normal modes is referred to as *normal mode analysis* (NMA).

The decomposition in normal modes is also applicable if the equation of motion, Eq. (3.43), is expanded to include an homogenous[4] friction term and a random force, analogous to the Langevin equation, Eq. (2.19). Each normal mode is given as one harmonically bound Langevin process with angular frequency $\omega_i = \sqrt{\theta_i}$ [120]. The introduction of noise and friction corresponds to the coupling with a heat bath. The variance of the normal mode position around the minimum, $\langle g_i^2(t) \rangle$, corresponds to a potential energy in a harmonic potential, $\omega_i^2 \langle g_i^2(t) \rangle /2$, with an angular frequency ω_i; note that the coordinates are mass-weighted. Due to the equipartition theorem, it is

$$\langle g_i^2(t) \rangle = \frac{k_B T}{\theta_i^2}. \qquad (3.45)$$

The above demonstration is based on the assumption of harmonic interaction potentials. Here, harmonicity is assumed for the Cartesian coordinates Eq. (3.41). If the potential is harmonic in any other set of coordinates, NMA can be analogously applied to the equations of motion expressed in the "linear" coordinate set. But normal modes can be a useful tool even for more general potentials [72, 120]; for sufficiently low temperatures the system will be close to a local

---

[4] We here assume that there is a unique friction with all atoms, *i.e.*, the friction is given by a number. A frequency dependent friction is discussed in [120].

potential minimum at position r_0. The potential around this minimum can be approximated by a harmonic expression. The matrix $\mathbf{H}$ is obtained as the Hessian of the potential, $V(r)$, i.e., the components of $\mathbf{H}$ are given as

$$H_{ij} = \frac{\partial^2 V(r_0)}{\partial r_i \partial r_j}, \tag{3.46}$$

and $\mathbf{H}$ characterizes the curvature of the potential at the local minimum in r_0. It must be stressed that this approximation is a local one: each local minimum has an individual Hessian. The accuracy of the approximation depends on the ratio between temperature and the leading correction to the harmonic approximation.

In MD simulation the normal modes of a particular local minimum can be obtained from the force field at a certain configuration r_0. A crucial prerequisite of the calculation is the position of the minimum, the configuration r_0. The position of the minimum is obtained by an energy minimization with different methods, like steepest descent, conjugate gradient, or limited Broyden-Fletcher-Goldfarb-Shanno [97].

3.5 Principal Component Analysis (PCA)

In an MD simulation with N particles, a large amount of high-dimensional data is generated; the trajectory contains $3N$ coordinates for each time step for which the positions are recorded. Multidimensional data analysis provides schemes that, depending on the quantities of interest, allow the amount of data to be reduced. One such scheme common in MD simulation is *principal component analysis*[5] (PCA) [122, 124–126].

In NMA, the motion of a system with approximately harmonic interactions is decomposed in different modes. In the case of an MD simulation of a large biomolecule at physiological temperature, the interactions are strongly anharmonic. However, there are choices of coordinates that allow different kinds of modes to be separated efficiently.

---
[5] The method was invented by the British mathematician Karl Pearson [121], the same who coined the term *random walk*. PCA has also been referred to as "essential dynamics" [122, 123], "molecule optimal dynamic coordinates" [124], and "quasi-harmonic analysis" [125].

3.5 Principal Component Analysis (PCA)

Assume an MD simulation yields a trajectory, $\mathbf{r}(t)$, of length T. The coordinates $r_i(t)$ are chosen such that the center of mass has zero velocity. The mass-weighted coordinates, $\mathbf{x} = \mathbf{M}\mathbf{r}$, have the components $x_i(t) = \sqrt{m_i}r_i(t)$. In the analysis of large biomolecules, we consider only the coordinates of the N particles that form the peptide or protein. Sometimes, the set of coordinates is further reduced, *e.g.*, to contain only the positions of the heavier atoms or the C_α-atoms. The surrounding water molecules and ions are assumed to follow considerably faster dynamics. The average value of the ith component of the mass-weighted position vector is denoted by $\langle x_i(\tau) \rangle_{\tau,T} = \bar{x}_i$, where τ is the variable over which the average is performed and T the length of the time interval in which the average is performed. PCA starts with the covariance matrix, sometimes also called the second moment matrix, $\mathbf{C}$, whose components are given as

$$C_{ij} = \langle (x_i(\tau) - \bar{x}_i)(x_j(\tau) - \bar{x}_j) \rangle_{\tau,T}. \tag{3.47}$$

The diagonal elements of the matrix $\mathbf{C}$ represent the fluctuations of the molecule. As is known from matrix algebra, the trace of a matrix, *i.e.*, the sum of the diagonal elements, is invariant with orthogonal basis set transformations. The trace of $\mathbf{C}$ is the total fluctuation of the molecule, which is independent of the choice of coordinates. The off-diagonal elements of $\mathbf{C}$ give the correlations between the coordinates. In contrast to the fluctuations, the correlations depend on the basis set chosen.

The covariance matrix is symmetric by construction. It is diagonalized by the orthogonal matrix $\mathbf{W}$, *i.e.*,

$$\mathbf{W}^T \mathbf{C} \mathbf{W} = \mathbf{\Lambda}, \tag{3.48}$$

where $\mathbf{\Lambda}$ has the components $\Lambda_{ij} = \delta_{ij}\lambda_i$. The matrix $\mathbf{W}$ represents a change from one orthonormal basis set to another. The eigenvalues of the covariance matrix, λ_i, are the fluctuations along the eigenvectors. We sort the eigenvalues in descending order, λ_0 being the largest eigenvalue. The coordinates, $\mathbf{q} = \mathbf{W}^T \mathbf{x}$, in which the covariance matrix is diagonal, are termed *principal components* (PC). By construction, the PCs are uncorrelated coordinates, but they are not statistically independent. The PCs are collective coordinates, *i.e.*, they involve multiple atoms. As in the case of normal modes, the low PCs are global coordinates involving many particles, while the high PCs are more localized. PCA allows those linear combinations of the original coordinates to be determined, which account for the strongest contributions to the overall internal motion. The time evolution of the component, $q_i(t)$, is the PC mode i.

The PCs are related to the original coordinates by a linear basis set transformation. The choice of original coordinates determines the class of basis sets which possibly can be determined with PCA. Therefore, it is of crucial importance to chose a proper coordinate set before launching the PCA. Instead of the Cartesian positions of a group of atoms, internal angles are also common starting coordinates for PCA [123, 127, 128]. In Cartesian coordinates, the removal of center of mass motion is obvious, whereas the removal of the molecular rotation is not uniquely defined and leads to some ambiguity [129–131].

3.6 Convergence

MD simulation trajectories do not correspond to real dynamics, but rather imitate real dynamics. As argued in Sec. 3.1, MD simulations attempt to capture the statistical properties of the systems investigated. Provided the trajectory is long enough to ensure a representative sampling of the configuration space, the ergodic hypothesis can be applied, and the ensemble properties can be derived from time averages. In this case, one says the trajectory is *converged*.

MD simulations are not the best way to sample the configuration space efficiently. Other methods, such as Monte Carlo simulations or umbrella sampling, are favorable in this respect. However, MD simulations do not only reproduce the equilibrium state but they also contain information about the time evolution of the system. That is, MD simulations are the best classical approximation for the *kinetics* of the system. One can increase the accuracy of the configuration space sampling by performing multiple simulations instead of a single long trajectory [132].

Real systems have a wide range of intrinsic time scales. Generally, larger systems with larger molecules tend to have longer intrinsic time scales. The simulation has to exceed the longest of these intrinsic time scales to reach the equilibrium state [133]. If the simulation time is shorter than any of the intrinsic

3.6 Convergence

time scales, the simulation is likely to provide a poor sampling of parts of the configuration space. The poor sampling may invalidate the statistical properties of the MD trajectory, as the properties cannot be considered representative of the system under investigation.

Usually, simulations are considered converged, if, *e.g.*, the fluctuations have reached a plateau value [103]. If the simulations are not converged, it is difficult to draw any conclusions from the trajectories. However, it has been shown repeatedly that the assumption of convergence is not justified in many cases [133–139]. Assessing the convergence of a given trajectory is a non-trivial task, "because it involves attempting to use what has been measured to deduce whether there is anything of importance which remains unmeasured" [139]. Strictly speaking, the convergence can be assessed only on the basis of an increased amount of data. However, various methods are used to test the convergence with the limited random sample obtained in an individual MD simulation.

A common test for convergence is the time dependence of the time-averaged MSD. If the trajectory is converged the time-averaged MSD reaches for long times a constant value. For multiple trajectories, the overlap, *i.e.* the similarity of the PC coordinates can quantify the convergence [136, 139]. Another measure for convergence is the cluster population in the configuration space. The configurations are cast into a finite set of reference structures (clusters). Then the trajectory is cut into pieces. If the various pieces of the trajectory exhibit the same cluster population as the whole trajectory, the simulation can be seen as converged [138]. This technique can also be used to compare different simulations [139]. A particular problem is the convergence in the context of PCA. The coordinate set obtained by PCA is unstable for unconverged trajectories [133, 136, 137]. An analytical treatment of Brownian diffusion reveals the cosine shape of the PC modes in an unconverged MD simulation. The cosine content allows the contribution of Brownian diffusion to the fluctuations to be measured [137]. However, even a zero cosine content does not safeguard convergence of the simulation.

After having introduced the methods that allow the time evolution of biomolecules to be simulated and studied, we now turn to examples of such computer simulations. We present in the next chapter results obtained from MD simulations of various biomolecules and discuss their thermodynamics and kinetics in some detail.

> *[...] if we were to name the most powerful assumption of all, which leads one on and on in an attempt to understand life, it is that all things are made of atoms, and that everything that living things do can be understood in terms of the jiggling and wiggling of atoms.*
>
> RICHARD P. FEYNMAN

CHAPTER 4

STUDY OF BIOMOLECULAR DYNAMICS

The results in this chapter have been partially published in T. Neusius, et al., Phys. Rev. Lett. **100**, *188103, cf. Ref. [140].*

4.1 Setup of the MD simulations

In order to study the thermal fluctuations of biomolecules and the kinetics that underlay the fluctuations, MD simulations of different peptides and a β-hairpin protein are performed. The MD simulations presented in the present thesis were performed by Isabella Daidone [141]. Three different models are used to simulate the dynamics of solvent water. The comparison between the three water models allows the influence of water on the kinetics of biomolecules in aqueous solution to be assessed. For the MD simulations, the GROMACS software package [97, 98] and the Gromos96 force field [99] are used. In the following, the details of the simulations are listed. Most of the numerical methods are explained in more detail in Chap. 3.

The $(GS)_nW$ peptides

The $(GS)_nW$ molecules are polypeptide chains formed of n (here $n = 2, 3, 5, 7$) repeated GS segments (G = glycine, S = serine) with a final tryptophan (W) residue. The $(GS)_nW$ peptides do not fold into a specific secondary structure.

Simulations of these peptides are performed with the LINCS algorithm [96] with an integration step of 2 fs. The data were saved every picosecond, the simulation length of each simulation is given in Tab. 4.2. The canonical ensemble is used, *i.e.*, the NVT ensemble with constant number of particles, N, constant volume, V, and constant temperature. The constant temperature, $T = 293\,\text{K}$, is established by the isokinetic thermostat [117].

Each peptide in extended conformation is placed in a rhombic dodecahedron box. The boundary of the boxes have at least a distance of ≈ 1.0 nm to the atoms of the peptide enclosed. The aqueous solution is modeled by the eSPC water model [112]. The liquid density is $55.32\,\text{mol/l}$ ($\approx 1\,\text{g/cm}^3$). The real space cut-off distance is set to 0.9 nm. Periodic boundary conditions are used and long-range interactions are treated by the particle mesh Ewald method [105] with a grid spacing of 0.12 nm and 4th-order B-spline interpolation.

The β-hairpin

MD simulations of the 14-residue amyloidogenic prion protein $H1$ peptide are performed using a similar setup as for the $(GS)_nW$ peptides. The native structure of the β-hairpin is a β-turn. Multiple folding and unfolding events can be observed during the simulation [141]. Thus, unlike the $(GS)_nW$ peptides, the β-hairpin has a secondary structure. The data are recorded all two picoseconds, the simulation time is $1\,\mu\text{s}$.

In contrast to the $(GS)_nW$ peptides, the β-hairpin simulation with explicit

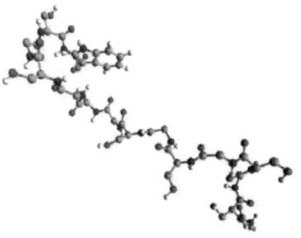

Figure 4.1: $(GS)_7W$ **peptide**. -

n	2	3	5	7
$(GS)_nW$	0.8	1.0	1.9	2.5

Figure 4.2: **Simulation lengths in units of microseconds (μs).** -

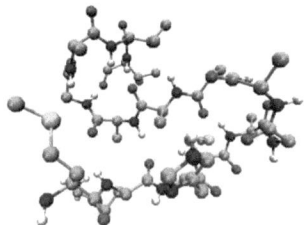

Figure 4.3: *β-**hairpin molecule**.* - 14-residue amyloidogenic prion protein $H1$ peptide.

solvent uses a truncated octahedron as simulation box. To check the role of the solvent dynamics, two alternative simulations are performed, besides the simulation in aqueous solution: one with an effective Langevin process imitating the solvent dynamics [97] and a simulation of the β-hairpin with implicit solvent using GB/SA [113, 114]. The Born radii are calculated using the fast asymptotic pairwise summation of [142]. The relevant Gromos96 parameters can be found in [143]. To increase the efficiency of the surface area calculation, a mimic based on the Born radii is used [144]. Further details of the β-hairpin simulations can be found in [141].

4.2 Thermal fluctuations in biomolecules

Due to the thermal agitation, the configuration, r, of a biomolecule exhibits internal fluctuations in equilibrium. These stationary fluctuations shall be studied in the following section employing PCA.

Let the coordinate system of the configuration space be such that the average position vector, r vanishes, $\langle r \rangle = 0$. The overall fluctuations of the Cartesian

54   STUDY OF BIOMOLECULAR DYNAMICS

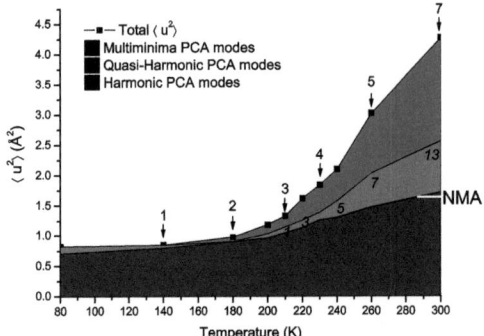

Figure 4.4: (Color online) **Total position fluctuation of a myoglobin protein as a function of temperature.** - For low temperatures the position fluctuation, in the figure denoted as $\langle u^2 \rangle$, increases linearly with the temperature T. Above 180 K, the fluctuations sharply increase, the so-called dynamical transition. The different colors correspond to different types of PCs, depending on the shape of the free energy profile along that mode (see page 56). The transition is mainly due to the appearance of non-harmonic PCs (red) above the transition temperature, some additional fluctuation is due to the quasi-harmonic modes (green), while the harmonic modes (blue) essentially contribute what is expected from the normal modes, indicated by NMA. The numbers at the curves give the quantity of PCs which belong to the different types of free energy profiles. Figure from A. Tournier [145].

positions[1] are

$$\langle r^2 \rangle = \sum_i \langle r_i^2 \rangle. \quad (4.1)$$

The thermal agitation is characterized by the temperature, T. Therefore, the overall fluctuation, $\langle r^2 \rangle$, increase with T. The fluctuations, $\langle r^2 \rangle$, of a myoglobin molecule as a function of T is illustrated in Fig. 4.4. The fluctuation exhibits a linear temperature dependence for low T, but above ≈ 180 K, $\langle r^2 \rangle$ sharply increases with the temperature, T: the molecule becomes much more flexible. The

---

[1] In this section, the same notation as in Chap. 3 is used. However, throughout this section, ordinary mean values, $\langle \cdot \rangle$, are to be understood as time averages over finite intervals, $\langle \cdot \rangle_{\tau,T}$, unless something else is explicitly indicated.

4.2 Thermal fluctuations in biomolecules

enhanced flexibility of the molecule is referred to as the "dynamical transition" or "glass transition". For details see the caption to Fig. 4.4.

In the case of a harmonic potential, the mass-weighted position fluctuations, $\langle x^2 \rangle = \langle r^T M r \rangle$, with the mass matrix, M, are proportional to the thermal energy, by virtue of the equipartition theorem. Therefore, in the following, the discussion is focused on the mass-weighted coordinates, x, rather than the Cartesian coordinates, r.

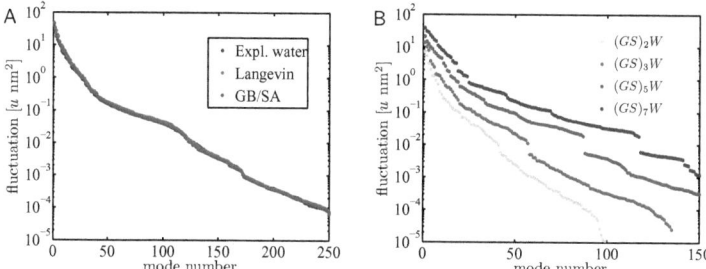

Figure 4.5: (Color online) **PCA eigenvalues.** - Fluctuations as measured by PCA eigenvalues. A: PCA eigenvalues of three different simulations of the β-hairpin. All simulation methods employed – explicit water, Langevin thermostat, GB/SA – exhibit the same spectrum of eigenvalues, i.e., the strength of the fluctuations seen in the explicit solvent simulation is reproduced by the implicit solvent methods (Langevin and GB/SA) with high accuracy. B: PCA spectrum of eigenvalues of the $(GS)_n W$ peptides ($n = 2, 3, 5,$ and 7): the longer chains are more flexible.

In any mass-weighted coordinate system, the total fluctuations are

$$\langle r^T M r \rangle = \langle x^2 \rangle = \langle q^2 \rangle = \sum_i \lambda_i, \quad (4.2)$$

where q is the configuration vector expressed in the basis of the PCs and the λ_i are the eigenvalues, as obtained from PCA. The expression on the right of Eq. (4.2) is the trace of the covariance matrix, Eq. (3.47). As the trace is invariant with orthogonal transformations, $\langle x^2 \rangle$ is independent of the basis set chosen. This independence reflects the physical nature of the fluctuations, $\langle x^2 \rangle$, which can be measured, e.g., by neutron scattering experiments [31].

The eigenvalues of the individual PC modes are illustrated in Fig. 4.5. The three different water models used for the β-hairpin exhibit a very similar spectrum of eigenvalues [Fig. 4.5 A]. The congruence of the three water models with respect to the eigenvalues demonstrates that the equilibrium fluctuations of the explicit water model are correctly reproduced by the implicit water methods. The spectrum of the $(GS)_nW$ peptides illustrates the increase of flexibility which comes along with an increasing chain length [Fig. 4.5 B]. The eigenvalues cover a range of five to six orders of magnitude. The large differences between the lowest and the highest eigenvalues accounts for the heterogeneity of the PC modes; the lowest PC modes contain much stronger fluctuations than the higher PC modes. Therefore, PCA is a useful tool in identifying low-dimensional subspaces in the configuration space, which contain a significant fraction of the overall fluctuations.

Potential of mean force

The MD simulations used in the present thesis were performed in the canonical ensemble, *i.e.*, the NVT ensemble. The thermodynamics of the canonical ensemble is given by the free energy[2]

$$F(T, V, N) = -k_B T \log \mathcal{Z}, \qquad (4.3)$$

where the partition function, $\mathcal{Z} = \int e^{-H/k_B T} d\Gamma$, follows from the Hamilton function of the system, H. The integral is over the phase space, Γ. Often, the state of the system is analyzed as a function of an appropriate reaction coordinate, y. Note that y may contain more than one free parameter. The free energy is calculated at a given value of the coordinate, $F_y(T, V, N)$, that is, the coordinate y acts as a constraint while the coordinates unaffected by the fixed value of the reaction coordinate are treated in the canonical ensemble. We stress that for general, curved coordinates the constrained free energy is not uniquely defined, but depends on the coordinates chosen. In the present thesis, only coordinates are used which can be obtained from the mass-weighted Cartesian coordinates by a linear, orthogonal transformation. Therefore, the constraint $y = $ const. is to be

---

[2] The free energy is sometimes referred to as *Helmholtz free energy*, in particular in the context of chemistry.

4.2 Thermal fluctuations in biomolecules

understood as the linear subspace spanned by y. Then, the integration over the remaining, orthogonal coordinates is uniquely defined.

In the thermodynamical equilibrium, the probability distribution obeys the Boltzmann statistics. Therefore, the probability distribution of the reaction coordinate, $\rho(y)$, determines the free energy

$$F_y(\mathcal{T}, V, N) = -k_B \mathcal{T} \log \rho(y). \tag{4.4}$$

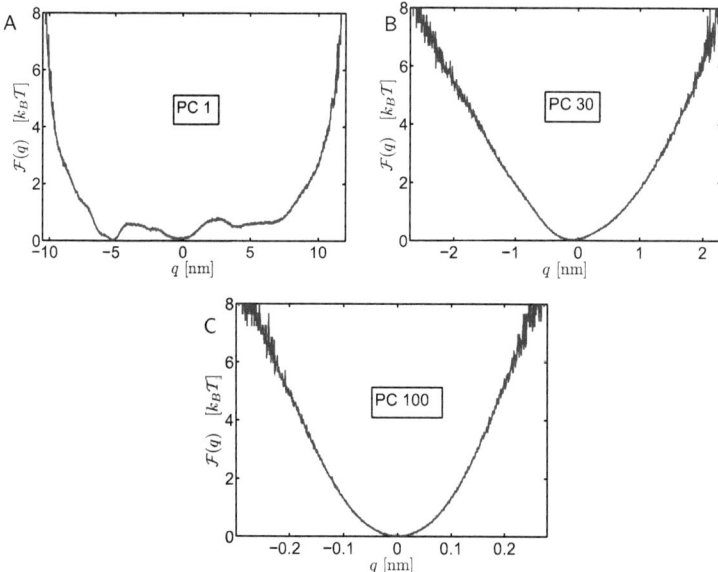

Figure 4.6: **Potential of Mean Force (PMF) of** $(GS)_5W$ - The potential of mean force is the free energy profile in which the dynamics of a single PC mode, q, evolves. It is obtained from a histogram, $h(q)$, of the PC mode as $F(q) = -k_B \mathcal{T} \log h(q)$. PC 1 of the $(GS)_5W$ peptide exhibits a multi-minima shape. PC 30 has an anharmonic PMF, but with a single local minimum (quasiharmonic). The PMF of PC 100 is harmonic.

In the following, we analyze the free energy profiles along the PCs, q_i. The free energy as a function of the constraint q_i is referred to as *potential of mean*

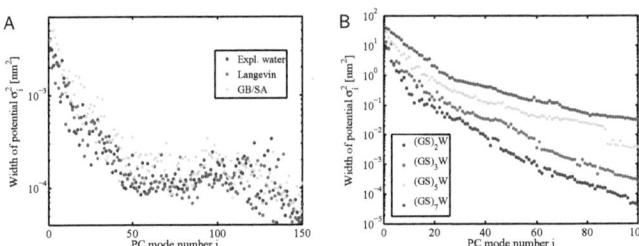

Figure 4.7: (Color online) **Fluctuations along PCs** - *The thermal fluctuations along the PCs of the three β-hairpin simulations (A) and the simulations of the four $(GS)_nW$ peptides (B), obtained from the variance of the PC modes. The low PCs exhibit the strongest fluctuations, as is expected from the definition of the PCA. All three simulation methods used for the β-hairpin exhibit a similar pattern of the fluctuations along the PCs. The PCs of the $(GS)_nW$ peptides fluctuate more for the longer chains (those with higher n). Note the similarity of the results illustrated in Fig. 4.5.*

force (PMF) [146]. The equilibrium distribution along the PC mode q_i can be obtained from the MD trajectory as a histogram, $h(q_i)$. The free energy is found, up to a constant, as

$$F(q_i) = -k_B\mathcal{T} \log h(q_i), \qquad (4.5)$$

in which the arguments $\mathcal{T}, V$, and N are skipped for simplicity.

The PC modes can be classified into three groups by virtue of their PMF: (i) PC modes with a harmonic PMF are called *harmonic* PCs, (ii) PC modes with a single local minimum in the PMF but which are not harmonic are referred to as *quasi harmonic* PCs, and (iii) those with various local minima, are designated as *multi-minima* PCs [145]. In Fig. 4.6, three examples of PMFs of the $(GS)_5W$ peptide are illustrated. The lowest PCs exhibit multi-minima shape, while the higher PCs follow in a very good approximation a parabola. The intermediate PCs feature a single minimum but are anharmonic. A similar behavior is found for the β-hairpin, both in explicit water [Fig. 4.8] and implicit water [Fig. 4.9]. Similar free energy profiles are found for the GB/SA simulation (not shown). The Langevin simulation and the GB/SA simulation of the β-hairpin reproduce the overall features of the explicit water simulation of the same molecule, as can be seen in Fig. 4.7, although the details of the free energy profiles look different.

4.2 Thermal fluctuations in biomolecules

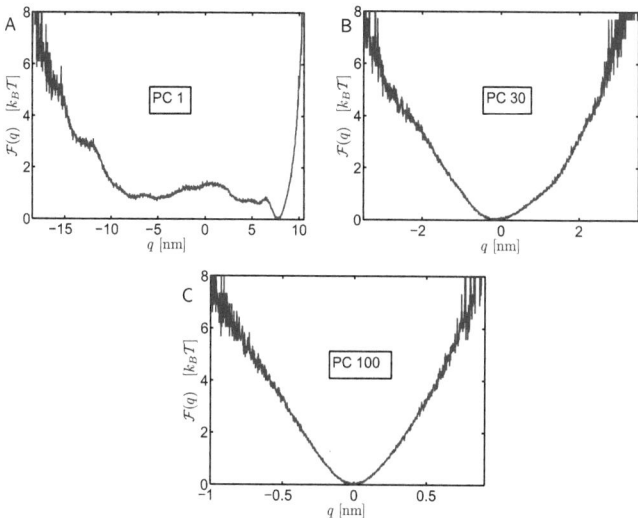

Figure 4.8: **Potential of Mean Force (PMF) of the β-hairpin simulation with explicit water.** - The potential of mean force is the free energy profile in which the dynamics of a single PC mode, q, evolves. It is obtained from a histogram of the PC mode as $F(q) = -k_B T \log h(q)$. The PMF is shown for PC 1, 30, and 100 of the β-hairpin in explicit solvent. A: Multi minima mode. B: Quasi-harmonic mode. C: Harmonic mode.

Only a small number, usually 10 to 20, of the PC exhibit a multi-minima PMF at physiological temperatures. However, as found by A. Tournier et al. [145], the overall fluctuations of a biomolecule at ambient temperature are mainly due to these few multi-minima PCs, see Fig. 4.4. The onset of anharmonicity in the lowest PCs has been identified as the origin of the dynamical transition of proteins at $T_g \approx 200$ K [145], illustrated in Fig. 4.4. PCA allows those degrees of freedom to be found, which contribute most to the overall fluctuations of the molecule. The observation of few coordinates which account for the majority of fluctuations, forms the basis for any attempt to reduce the number of degrees of freedom in the modeling of biomolecules, cf. Sec. 5.1.

As is illustrated in Fig. 4.7, the lower PCs have a broader PMF. The width of the potential is proportional to the fluctuations along the reaction coordinate.

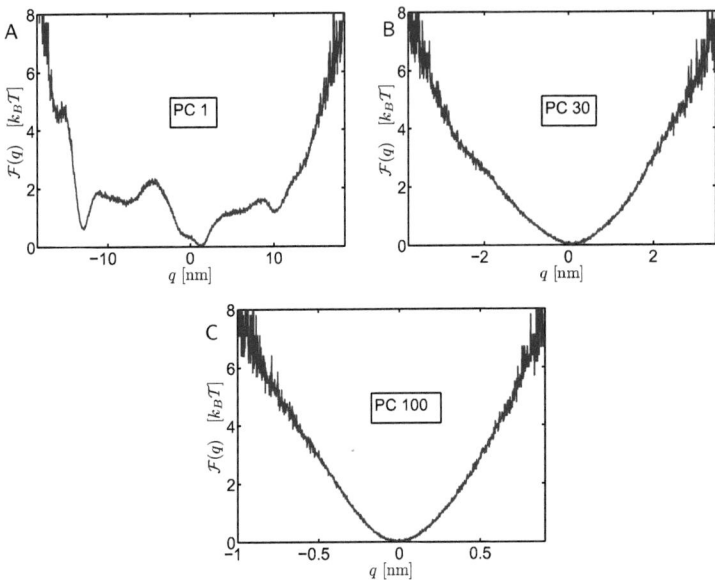

Figure 4.9: **Potential of Mean Force (PMF) of the β-hairpin simulation with Langevin thermostat** - *The potential of mean force is the free energy profile in which the dynamics of a single PC mode, q, evolves. It is obtained from a histogram of the PC mode as $F(q) = -k_B T \log h(q)$. The PMF is shown for PC 1, 30, and 100 of the β-hairpin with implicit solvent modeled by a Langevin thermostat.* **A**: *Multi-minima mode.* **B**: *Quasi-harmonic mode.* **C**: *Harmonic mode.*

The three simulation methods used for the β-hairpin provide similar widths of the PMFs, Fig. 4.7 A. Thus, the Langevin and GB/SA simulations reproduce the fluctuations as seen in the explicit water simulation. The agreement between the explicit and implicit water methods corroborates the findings in the context of the spectrum of eigenvalues, *i.e.*, the free energy profiles and the total fluctuation are well reproduced by all three water simulation techniques. The configurations of the longer peptides populate a larger volume in the configuration space, Fig. 4.7 B. The larger configuration volume can be understood as an enhanced flexibility of the chain. The decrease of the fluctuations with the mode number appears to be more regular for the $(GS)_nW$ peptides, relative to the pattern of the β-hairpin simulation.

The kinetics along a reaction coordinate are not fully described by the free energy. This is due to the fact, that the projection to the subspace (or, more generally, the manifold) spanned by the possible values of the reaction coordinate, may lead to time correlations or memory effects, as follows from Zwanzig's projection operator approach (cf. Sec. 5.1). Only on sufficiently long time scales, when all memory effects have decayed, the kinetics follow directly from the free energy profile. Therefore, in general, the free energy profiles are insufficient to understand the kinetic behavior of a molecule.

Participation ratio

Both, the PC modes and normal modes, are collective coordinates, *i.e.*, they are linear combinations of the original coordinates, the components of $\boldsymbol{x}$. As PCA is based on a linear coordinate transformation, the choice of original coordinates constrains the potential PCs to those basis sets, which can be yield by a linear basis transformation of the original coordinates. In the original coordinates, the PC k has the components W_{ik}, *i.e.*, PC k is given as the kth column of the matrix $\mathbf{W}$ which diagonalizes the covariance matrix, $\mathbf{C}$, cf. Eq. (3.48). To quantify the delocalization of PC k, the participation ratio is introduced as

$$\eta_k = \sum_{i=1}^{3N-6} W_{ik}^4. \tag{4.6}$$

If the PC k involves just one atom, the PC eigenvector k has the components $W_{ik} = \delta_{ik}$, leading to $\eta_k = 1$. A collective motion that involves all atoms homogeneously has the eigenvector components $W_{ik} = 1/\sqrt{3N-6}$, as follows from the normalization of the basis vectors. The participation factor then equals $\eta_k = 1/(3N-6)$. Small participation ratios correspond to delocalized PCs, whereas values close to one indicate strong localization. The reciprocal value of η_k corresponds to the average number of degrees of freedom that are involved in the PC, the *participation number*. The participation number of the β-hairpin simulations is illustrated in Fig. 4.10 A. All three simulation methods exhibit modes that are similarly delocalized, albeit with strong irregularities. The low PCs are spread over the whole molecule, whereas the higher PCs tend to be more localized. The participation number of the $(GS)_nW$ peptides has a similar dependence on the mode number, *i.e.*, the lower PCs are strongly delocalized; with increasing mode number the participation number decreases.

Anharmonicity

If the interaction potential of a molecule is purely harmonic, as in Eq. (3.41), the covariance matrix can be expressed in terms of the matrix **U**, which transforms the original coordinates to the normal modes [Eq. (3.43)], and the normal mode eigenvalues θ_k

$$C_{ij} = \sum_k U_{ik} U_{jk} \theta_k, \qquad (4.7)$$

where the U_{ij} are the components of the matrix **U**. From the definition of PCA it follows that the matrix **U** solves the PCA eigenvalue equation, Eq. (3.48), for harmonic interaction potentials. Hence, the matrix that transforms the original coordinates to the PCs is identical to the normal mode matrix, *i.e.* **W** = **U**. Therefore, if the interaction potential is purely harmonic, the normal modes, $g_i(t)$,

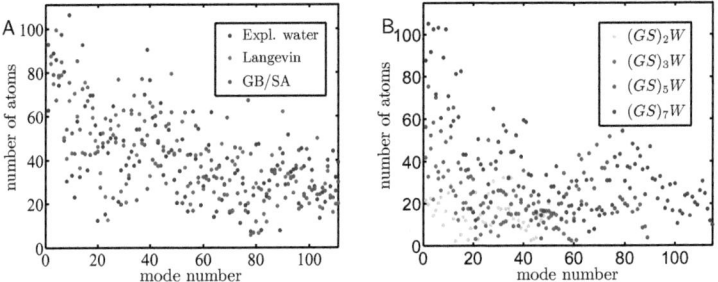

Figure 4.10: *(Color online)* **Participation number** - A: *The participation number of PCs of the different simulations of the β-hairpin (111 atoms). The PCs involve a large number of atoms and are clearly non-local. In particular the low PC modes are non-local. The participation number is not affected by the different simulation methods for the water dynamics.* B: *The participation number of the $(GS)_nW$ ($n = 2, 3, 5,$ and 7) molecules, which consist of 50, 63, 89, and 115 atoms, respectively. The nonlocality is clearly visible and is more pronounced for the lower PCs.*

4.2 Thermal fluctuations in biomolecules

collapse with the PC modes, $q_i(t)$, and the eigenvalues of the normal modes, θ_i, are identical to those obtained by PCA, *i.e.* $\theta_i = \lambda_i$.

When describing the fluctuations around a minimum of an anharmonic potential, the difference between the fluctuations as seen in the simulation and the normal mode contribution – which is an approximation for low temperatures – can quantify, how much is contributed in excess of the harmonic contribution. That is, the fraction of the fluctuations due to the harmonic normal mode approximation allows the anharmonicity to be estimated [24]. In the situation of anharmonic interaction potentials, PCA leads to different coordinates than NMA, *i.e.*, $\mathbf{U} \neq \mathbf{W}$. The contribution of the normal modes to PC mode i are obtained by projecting the normal mode fluctuations to the one-dimensional subspace spanned by PC mode i. Upon projection, the fluctuation along the principal component i, which would be expected from the normal mode approximation, is

$$\zeta_i = \sum_{jk} (W_{ji} U_{jk})^2 \theta_k. \tag{4.8}$$

The anharmonicity of PC mode i is defined in [24] as

$$\varsigma_i = \sqrt{\frac{\lambda_i}{\zeta_i}}, \tag{4.9}$$

i.e., as the total fluctuations divided by the fluctuations contributed by the normal modes. In a system with purely harmonic interactions, the anharmonicity factor, ς_i, equals one, otherwise it is larger than one.

It is assumed in the above argument that the mean configuration $\langle \boldsymbol{r} \rangle$ is given by the configuration at a unique potential minimum. If there are multiple local minima, NMA can be applied to each of the local minima, leaving the definition of ς_i in Eq. (4.9) inconclusive. Therefore, a uniquely defined measure of anharmonicity is required instead of the anharmonicity factor, ς_i.

In order to quantify the deviation of the PMF from a harmonic potential without referring to NMA, one can estimate the difference between the histogram, $h(q_i)$, and a Gaussian fit, $n_{\mu,\sigma}(q)$, to the histogram in an adequate norm of a

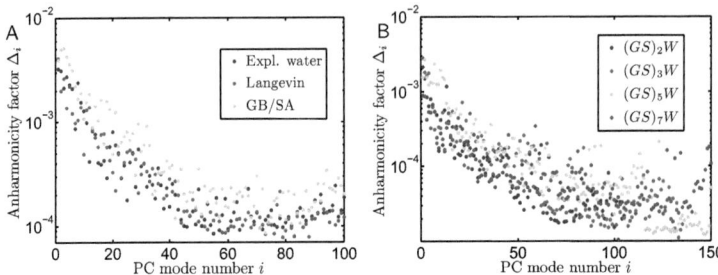

Figure 4.11: (Color online) **Anharmonicity degree, Δ_i for single PCs** - For each PC mode the best fit of a Gaussian to the underlying probability distribution function is performed. The anharmonicity degree is calculated using Eq. (4.10). A: Anharmonicity degree Δ_i of the β-hairpin. The anharmonicity is similar for the explicit water model, the Langevin simulation, and the GB/SA approach. The PMF of the β-hairpin exhibits a strong anharmonicity for the PC modes above $i = 40$. B: The anharmonicity of the PMF of the $(GS)_nW$ peptides ($n = 2, 3, 5,$ and 7). The PMFs of all peptides exhibit strong anharmonicity for the lower modes. There is a trend of Δ_i to decrease with increasing mode number.

function space, e.g., as the L^2-norm[3]

$$\Delta_i = \|h - n_{\mu,\sigma}\|_{L^2}^2 = \sum_i (s_{i+1} - s_i)|h(s_i) - n_{\mu,\sigma}(s_i)|^2. \qquad (4.10)$$

The *anharmonicity degree*, Δ_i, equals zero for a harmonic PMF. The more the PMF deviates from the harmonic one, the higher is Δ_i. The anharmonicity degree is illustrated for the β-hairpin and $(GS)_nW$ simulations in Fig. 4.11 as a function of the mode number. The lowest modes are strongly anharmonic. With increasing mode number, there is a tendency of Δ_i to decrease. Thus, the higher modes are more likely to exhibit a PMF that is approximately harmonic. However, the anharmonicity degree exhibits an irregular shape and scatters irregularly around the trend. All water models employed in the simulations of the β-hairpin lead to similar anharmonicity in the PMF.

---

[3] The functional space $L^2(V)$ is defined as the set of functions, $f : V \to \mathbb{R}$, which are square integrable. That is, for all $f \in L^2(V)$ the L^2-norm is finite, i.e., $\int_V |f(x)|^2 dx < \infty$.

4.3 Thermal kinetics of biomolecules

The quantities discussed so far can be obtained from the equilibrium distribution in the configuration space. In this sense, quantities like the PMF or the fluctuation along a given mode are stationary and do not represent kinetic aspects of the molecular behavior. In order to describe the kinetic behavior of the fluctuations we analyze the MSD of the PCs.

The MSD for the discrete data $q_n = q(n\Delta t)$, with Δt being the resolution of the trajectory, is obtained as

$$\left\langle \Delta x^2(t) \right\rangle = \left\langle \Delta x^2(n\Delta t) \right\rangle = \frac{1}{N-n} \sum_{k=1}^{N-n} (q_{k+n} - q_k)^2, \tag{4.11}$$

where N is the total number of frames of the trajectory, the length of the simulation being $T = N\Delta t$. The time averaging procedure in Eq. (4.11) is valid for $t \ll T$. For t-values too close to T, the trajectory is statistically not significant to perform the time average in Eq. (4.11).

Fig. 4.12 illustrates the time-averaged MSD of the individual PC modes of the β-hairpin simulations. For the explicit water simulation, Fig. 4.12 A, the PC 1 exhibits a power-law behavior extending from 2 ps up to 100 ns. The simulation length, $T = 1\,\mu$s, is too short to observe a significant saturation of the MSD, which is displayed in Fig. 4.12 A only up to $t = 0.1\,\mu$s, as the MSD can be obtained only for the time scale $t \ll T$. The exponent of the power law is in the range of ≈ 0.5, but the value is slightly different for the PC 2 and PC 3. The PCs 10, 20, and 30 also exhibit a power-law behavior for the short time scales but they reach more quickly the saturation. A subdiffusive MSD is also found in the Langevin simulation, Fig. 4.12 B, and the GB/SA simulation, Fig. 4.12 C. However, the MSD of the implicit water simulations provides power-law exponents ≈ 0.4 and ≈ 0.3, respectively, which are smaller than the explicit water simulation. The difference between the subdiffusion exponents of the explicit water simulation and the GB/SA method prompts to potential inaccuracies of the implicit water simulations when reproducing the kinetics of the β-hairpin.

The Langevin simulation mimics the effect of the solvent with an uncorrelated, white noise and a friction term. The GB/SA simply modifies the electrostatic

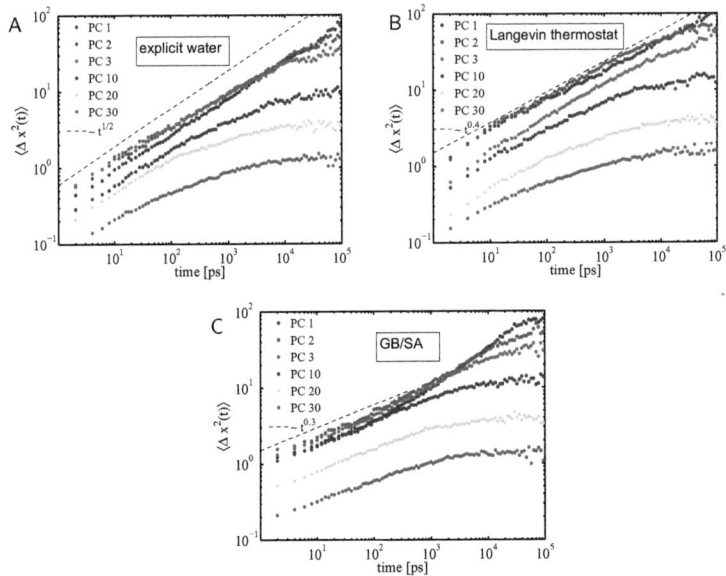

Figure 4.12: (Color online) **MSD of β-hairpin simulations with different water models** - The time-averaged MSD of the β-hairpin for various PCs. **A**: The explicit water simulation reveals a subdiffusive MSD for the lowest PCs. The exponent of the power law is ≈ 0.5. The higher PCs exhibit a quicker saturation. However, for short times, the subdiffusive regime displays MSD exponents similar to those for the low PCs. **B**: The Langevin dynamics simulation reproduces essentially the subdiffusive behavior found in the MSD of the explicit solvent simulation (**A**). However, the power law exponent is slightly smaller and equals ≈ 0.4 for PC 1. **C**: The GB/SA implicit solvent simulation exhibits subdiffusion in the MSD with an exponent ≈ 0.3 in the range of 2 ps to 1 ns. The low PC modes exhibit an increased MSD on a time scale of 1 to 10 ns – a behavior not found for any of the other simulation methods.

interaction potentials to imitate the presence of water, thereby ignoring friction and exchange of angular momentum.

The removal of the water dynamics can be expressed as a projection to a subspace of the configuration space, which is spanned by the internal coordinates of the β-hairpin. As follows from Zwanzig's projection formalism, the projection

4.3 Thermal kinetics of biomolecules

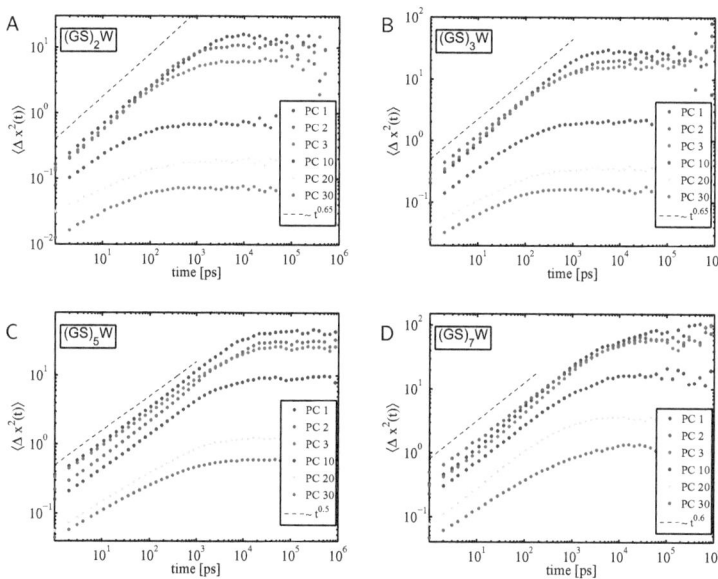

Figure 4.13: (Color online) **MSD of** $(GS)_nW$ **peptides** - The time-averaged MSD of the $(GS)_nW$ peptides for various PCs. The figures **A** to **D** correspond to $n = 2, 3, 5,$ and 7, respectively. For the lowest PCs, all peptides exhibit a subdiffusive MSD over at least three orders of magnitude in the time domain. An equilibrium is reached for the longest time scales when the MSD, $\langle \Delta x^2(t) \rangle$, saturates and does not further increase with increasing time, t.

gives rise to memory effects, which are not reproduced by the implicit solvent methods. Zwanzig's projection approach is discussed in more detail in Sec. 5.1. Surprisingly, the subdiffusive behavior, which is in a sense a memory effect, cf. Eq. (2.27), is more pronounced for the implicit solvent simulations.

The MSD of the $(GS)_nW$ peptide simulations is given in Fig. 4.13. Subdiffusive behavior is found in all simulations. The exponents of PC 1 are in the range from 0.65 [$(GS)_2W$ in **A** and $(GS)_3W$ in **B**, respectively] to ≈ 0.5 for the $(GS)_5W$ peptide (**C**). All simulations reach clearly the saturation plateau. The MSD saturates after 10^3 to 10^4 ps for the smallest peptide, $(GS)_2W$, and around

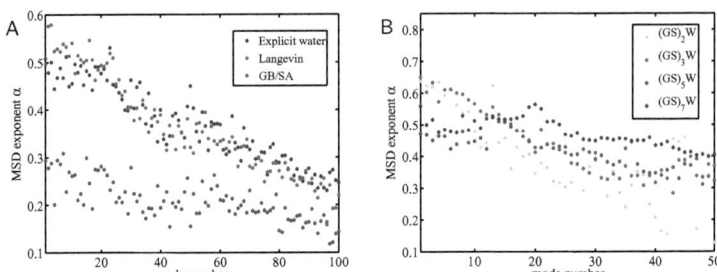

Figure 4.14: (Color online) **MSD exponents** *- The MSD exponents are obtained with a least-squares fit to the MSD in a double-logarithmic scale. The fit is performed in the time window from 1 ps to 10 ps to avoid an influence of the cross over to the saturation plateau.* **A:** *MSD exponents of the different β-hairpin simulations. The MSD in the PCs of the β-hairpin, simulated with explicit water, exhibit exponents ≤ 0.6, indicative of subdiffusion (blue dots). The exponent has a tendency to decrease with the mode number. The subdiffusion exponents are reproduced by the Langevin simulation (red dots). In contrast, the implicit water GB/SA simulation exhibits smaller MSD exponents (green dots), the exponents found in the GB/SA simulation are approximately half as big as those of the explicit water simulation.* **B:** *The $(GS)_n W$ peptides exhibit a subdiffusive MSD for all n = 2, 3, 5, and 7. The exponents have a tendency to decrease, a behavior more expressed for the smaller peptides.*

10^5 ps for the largest peptide, $(GS)_7 W$. Even in the case of $(GS)_7 W$, the MSD of PC 1 reaches a plateau. This gives evidence that the trajectory can be considered as converged with respect to the MSD calculation.

A linear least-squares fit to the logarithm of the data is performed in the time range 1 to 10 ps to obtain the exponents of the MSD power-law behavior. The fit is restricted to the shortest time scale to avoid an influence of the cross-over to the saturation plateau. The exponents can be found in Fig. 4.14 for both the β-hairpin simulations and the $(GS)_n W$ peptides. From Fig. 4.14 A it can be concluded, that the subdiffusive MSD obtained from the Langevin simulation reproduces the explicit water dynamics reasonably – at least in the range 1 to 10 ps, in which the fit is performed. In contrast, the GB/SA simulation fails to reproduce the kinetic behavior and underestimates the MSD exponents by a factor of two.

First passage times

The first passage time of a random walker is the time the walker needs to escape from a given volume. The first passage time distribution (FPTD) is a second kinetic quantity analyzed in this section, besides the MSD.

The asymptotic behavior of the FPTD is obtained by mapping the dynamics along the principal coordinates, q_i, onto a two-state process. For a given PC i, the range of q_i values observed during the simulation is partitioned into two parts, $q_i \geqslant a$ and $q_i < a$. For multi-minima PC modes, the value a is chosen to represent the location of the highest free energy barrier. For the harmonic and quasi-harmonic modes, by symmetry, a is chosen at the position of the minimum of the free energy profile. The results were found to be independent of the precise location of a. The PC mode $q_i(t)$ is then mapped to a binary series $b_i(t) \in \{0, 1\}$, such that $b_i(t) = 0$ if $q_i(t) \leqslant a$ and $b_i(t) = 1$ if $q_i(t) < a$. Then, a histogram of the times spent in state 0 and 1 are obtained. As the statistics are poor for the long-time behavior, the FPTD is convoluted piecewise with filters of different sizes. The piecewise filtering is required as the distribution covers four orders of magnitude.

The FPTD of the β-hairpin is illustrated in Fig. 4.15 and the FPTD of various $(GS)_n W$ peptides is shown in Fig. 4.16. For long times t, the FPTD exhibits a power-law behavior, $w(t) \sim t^{-1-\beta}$, again extending to the 10 ns range. The exponent β lies in the range of 0.5 to 0.6 for the lowest modes. A distribution with $w(t) \sim t^{-1-\beta}$ and $0 < \beta < 1$ has no mean value. It represents strong, long-lasting correlations in the kinetic behavior.

The power-law tail of the FPTD is a consequence of the memory effects that dominate the dynamics. Barrier-crossing processes in a one-dimensional Langevin dynamics lead to an exponential FPTD, as follows from Kramer's escape theory [69]. The dynamics along the PC i, are not uniquely determined by the free energy profile, *i.e.* the PMF. On the contrary, the dynamics are strongly influenced by the dynamics in the orthogonal PCs $j \neq i$, which contribute correlations to the PC i and give rise to memory effects up to the time scale of ≈ 10 ps.

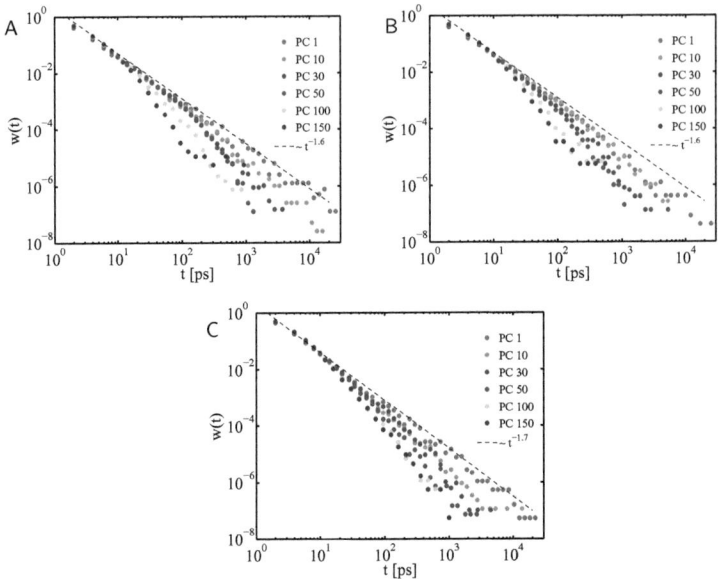

Figure 4.15: *(Color online)* **First passage time distribution (FPTD)**, $w(t)$, **of the β-hairpin.** - The FPTD is obtained by projecting time series of each PC onto a two-state space. For PCs with multi-minima potentials of mean force, the two states are defined by dividing the PC coordinate, q, into the two parts either side of the position of the highest barrier in the potential of mean force. For the higher PCs, which have only a single minimum, by symmetry the space is divided at the position of the minimum. The results are found to be independent of the precise location of the partition. In order to improve statistics at long times the data are piecewise convoluted with filters of different sizes. A: Explicit water simulation, B: Langevin dynamics simulation, C: GB/SA simulation.

Convergence

An important issue for the discussion of the MD simulations is the question as to whether the trajectories are long enough to sample the configuration space sufficiently. That is, has the system reached a state of thermal equilibrium on the time scale of the simulation? As pointed out in Sec. 3.6, this question is difficult to assess: we do not have multiple trajectories of the same system at our disposal

4.3 Thermal kinetics of biomolecules

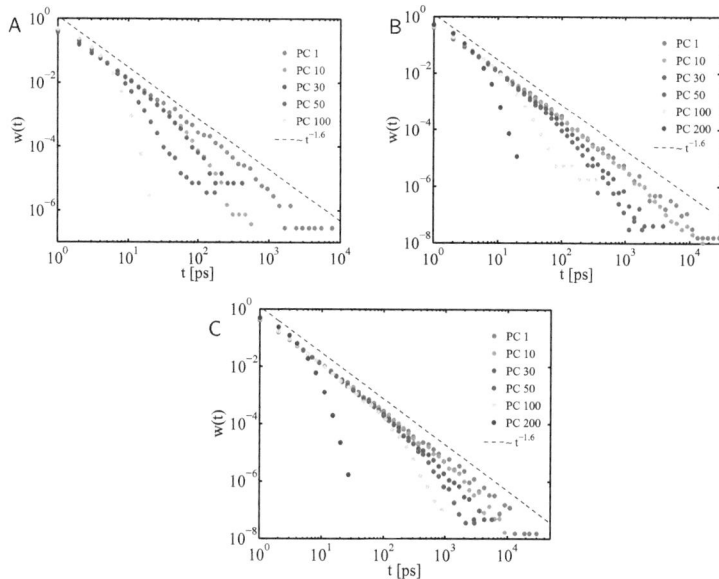

Figure 4.16: (Color online) **First passage time distribution (FPTD)**, $w(t)$, **of the** $(GS)_nW$ **peptides**. - The FPTD is obtained by projecting time series of each PC onto a two-state space. For PCs with multi-minima potentials of mean force, i.e. the lowest, the two states are defined by dividing the PC coordinate, q, into the two parts either side of the position of the highest barrier in the potential of mean force. For the higher PCs, which have only a single minimum, by symmetry the space is divided at the position of the minimum. The results are found to be independent of the precise location of the partition. In order to improve statistics at long times the data are piecewise convoluted with filters of different sizes. A: $(GS)_2W$, B: $(GS)_5W$, C: $(GS)_7W$.

to compare with. The MSDs of the PC 1 in Fig. 4.13 A-D exhibit a plateau at least at the time scale $t \approx T/10$, *i.e.* one order of magnitude below the simulation length, T. As the constant saturation plateau is an equilibrium effect, all of the $(GS)_nW$ simulations can be considered as converged with respect to the MSD.

An analysis of the PMFs affirms the convergence of the $(GS)_nW$ simulations. When cutting the trajectory into two pieces, a first and a second half, two histograms, from the first and the second half, are calculated successively, h_1 and

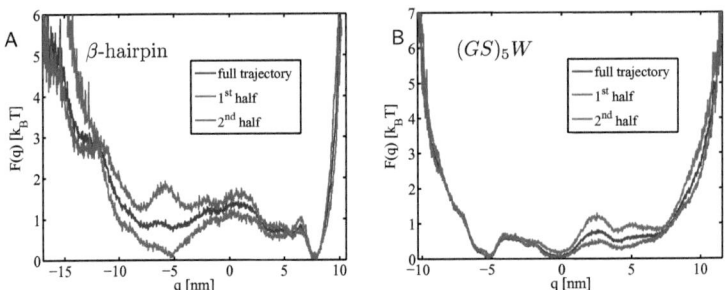

Figure 4.17: (Color online) **Free energy profile (PMF) for two halves of simulation** - Histograms of the PC mode 1 are obtained from the first and the second half of the simulation, independently. **A:** The β-hairpin PMF (explicit water simulation) exhibits substantial differences between the first (red) and the second half (magenta) of the simulation. The difference indicates the poor sampling of the configuration space, i.e., the trajectory is unconverged. **B:** The two halves of the $(GS)_5W$ peptide simulation exhibit the same structure and are in general close to each other. The sampling is much better relative to A, albeit not perfect.

h_2, respectively. In Fig. 4.17, the free energy profile (i.e. the PMF) for the PC mode 1 is illustrated for the β-hairpin simulation with explicit water (**A**) and the $(GS)_5W$ peptide. The β-hairpin PMF exhibits strong differences between the profile obtained from the first half and the one obtained from the second half, which is due to poor sampling. In contrast, the two free energy profiles of the first PC mode of $(GS)_5W$ nearly collapse indicating a rather good convergence. We calculate the L^2-difference between the two histograms, h_1 and h_2, that is

$$\Lambda_i = \|h_1 - h_2\|_{L^2}^2 = \sum_i (s_{i+1} - s_i)|h_1(s_i) - h_2(s_i)|^2. \qquad (4.12)$$

The convergence factor, Λ_i, is illustrated in Fig. 4.18 for the β-hairpin simulations (**A**) and the $(GS)_nW$ peptides (**B**). The differences between the first and the second half of the simulations are larger for the lower PCs containing the diffusive contribution.

In contrast to the $(GS)_nW$ peptides, the MSD of the β-hairpin PC 1 in Fig. 4.12 **A** does not reach a constant plateau. Therefore, the β-hairpin simulation is not fully converged on the time scale of simulation length, $T = 1\,\mu\text{s}$. This is

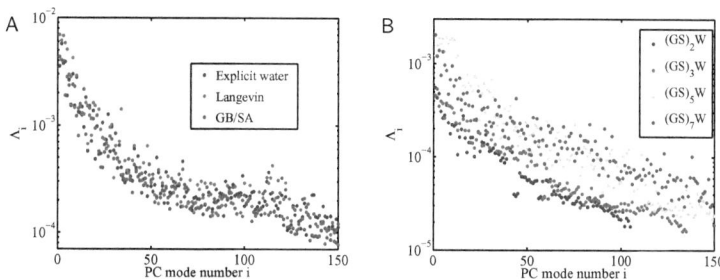

Figure 4.18: (Color online) **Convergence factor.** - Histograms of the PC coordinates are obtained from the first and the second half of the simulation, independently. The L^2-norm of the difference between the two histograms is given as the convergence factor, Λ_i, for every PC i. **A:** β-hairpin simulation, **B:** $(GS)_nW$ peptides.

confirmed by the Λ_i as illustrated in Fig. 4.18. The lowest Λ_i are approximately a factor ten larger for the β-hairpin in Fig. 4.12 A than the values of $(GS)_7W$ (B).

4.4 Conclusion

MD simulations of a β-hairpin molecule have been studied on a time scale of 1 μs. Three different methods to imitate the dynamics of solvent water have been used: the eSPC explicit water model, a Langevin thermostat and a method using the generalized Born equation. Also, MD simulations of four $(GS)_nW$ peptides with $n = 2, 3, 5,$ and 7 have been analyzed on the microsecond time scale. A PCA has been performed for each of the above-mentioned systems. The PMF of the low PCs exhibits strong anharmonicities in all systems. The lowest PCs are delocalized, general motions which involve a large fraction of atoms in the molecule. The PMF and the delocalization of the PCs exhibit the same features irrespective of the water model used.

The kinetics of the molecules is characterized in terms of the MSD of the individual PC modes. The MSD of the explicit solvent simulations exhibits a subdiffusive pattern in the time range from 1 ps to the range of nanoseconds.

The $(GS)_nW$ peptides do not fold into a unique secondary structure. The observation of a subdiffusive MSD in the PC modes of these peptides proves that a complex, secondary structure as seen by proteins is not a requirement for fractional diffusion in the internal coordinates of molecules. Therefore, the subdiffusivity appears as a widespread, general feature of biomolecular fluctuations.

The MSD found in the β-hairpin for the Langevin thermostat is similar to the subdiffusive MSD of the explicit water simulation. The Langevin thermostat mimics the effect of the solvent water dynamics. It does not include, however, any kind of memory effects, potentially arising from the projection of the dynamics to the subspace of the internal coordinates of the molecule, cf. Sec. 5.1. Therefore, the coincidence of the results of the Langevin thermostat and the simulation with the eSPC water model indicates that on the time scale of and above 1 ps, memory effects due to the projection to the internal molecule coordinates are negligible. That is, the detailed water dynamics can be ignored in the analysis of the internal subdiffusion, at least on the time scales of 1 ps and above.

In contrast, the GB/SA simulation exhibits a subdiffusive MSD which is different from the behavior found in the explicit water simulation. The exponents of the MSD power law are substantially smaller in the GB/SA simulations than the exponents found in the explicit water simulation. The difference between the simulation with explicit water and the simulation based on the generalized Born equation prompts to inaccuracies of the kinetics, as found with the GB/SA simulations.

When projecting the dynamics to a single PC mode, the dynamics is characterized by strong memory effects on time scales up to at least 10 ns. The memory effects are illustrated by the FPTD, which exhibits a power-law decay.

The $(GS)_nW$ peptide simulations have reached an equilibrium on the timescale of microseconds, whereas the β-hairpin simulation has poorly converged on the time scale of the simulation, $T = \mu$s. In any case, the subdiffusive MSD found in all of the systems analyzed is not an out-of-equilibrium effect but can be considered to be part of the stationary kinetics at ambient temperature.

In the next chapter, we address the question as to which mechanism gives rise to the subdiffusive MSD found in the MD simulation presented here. Various models are considered and compared with the MD simulation results obtained.

> Nihil certi habemus in nostra scientia, nisi nostram mathematicam.
>
> NICOLAUS VON KUES

CHAPTER 5
MODELING SUBDIFFUSION

*The results in this chapter have been partially published in T. Neusius, et al., Phys. Rev. Lett. **100**, 188103, cf. Ref. [140] and T. Neusius, et al., Phys. Rev. E **80**, 011109 cf. Ref. [147].*

The presence of subdiffusive kinetics in the internal coordinates of biomolecules raise the question as to the underlying mechanism. Several models can account for subdiffusion, some of which were discussed as candidates for biomolecular fluctuations. In the following chapter, some of these models are reviewed. It is demonstrated how the models compare with the experimental and simulation results. In particular, the continuous time random walk (CTRW) and the diffusion on fractal geometries are subjects of the present thesis.

First, we discuss briefly the projection approach of R. Zwanzig as described in [69]. Zwanzig's formalism allows the Hamiltonian equations to be written in a subspace of reduced dimensionality. Correlations and memory effects are the price one has to pay for the reduced number of degrees of freedom.

5.1 Zwanzig's projection formalism

Many systems of physical interest possess so large a number of degrees of freedom that it is impossible to follow simultaneously all of them in an experiment or in theoretical considerations. And even if it is possible, it may be more efficient to omit irrelevant details and describe the system in terms of the *relevant* aspects, as is done, for example, when Brownian motion is modeled by the Langevin equation. Here, we aim at establishing a framework, in which the system can be approximately described with reasonable effort. Furthermore, it is of high importance to estimate the accuracy of the approximation. Zwanzig's projection operator approach allows a Hamiltonian system to be described in a subspace of the phase space. The subspace is spanned by the set of relevant coordinates; irrelevant coordinates are neglected.

Assume a Hamiltonian system with coordinates q_i and momenta p_i and the Hamilton function $H(\boldsymbol{q},\boldsymbol{p})$. The phase space, Γ, is spanned by the q_i and p_i. Let $A(\boldsymbol{q},\boldsymbol{p},t)$ be a dynamical variable depending on the phase space coordinates, the momenta, and time the t. The time evolution of A is given by the Liouville equation

$$\frac{\partial}{\partial t} A = \mathbf{L} A, \qquad (5.1)$$

in which the Liouville operator is used, defined as

$$\mathbf{L} = \sum_i \left(\frac{\partial H}{\partial p_i} \frac{\partial}{\partial q_i} - \frac{\partial H}{\partial q_i} \frac{\partial}{\partial p_i} \right). \qquad (5.2)$$

Note that the Liouville equation is a differential equation. That is, the time derivative of A at time t^* depends exclusively on the values of A at that time t^*. There are no time correlations present, *i.e.*, the values of A in the past $t < t^*$ or in the future $t > t^*$ do not influence the time derivative in t^*. This property is referred to as the *Markov property* and a system is Markovian if it complies with that property. All Hamiltonian systems are Markovian, as they all can be described by Eq. (5.1). In what follows, we do assume Eq. (5.1) to be valid, *i.e.*, we assume the system to be Markovian. However, we do not explicitly use the form of the Liouville operator which follows from the Hamilton function. Therefore, the results of this section are valid for systems whose time evolution can be described by an equation of the form Eq. (5.1).

The dynamical variable A is a function on the phase space, $A : \Gamma \to \mathbb{R}$. We assume A to be square integrable, *i.e.*, the integral $\int_\Gamma |A|^2 \mathrm{d}\boldsymbol{q}\mathrm{d}\boldsymbol{p}$ is finite. The space of all square integrable functions is denoted as $L^2(\Gamma)$ and is a Hilbert space with the scalar product

$$\langle A | B \rangle = \int_\Gamma AB \mathrm{d}\boldsymbol{q}\mathrm{d}\boldsymbol{p}. \qquad (5.3)$$

The Liouville operator maps the space onto inself, $\mathbf{L} : L^2(\Gamma) \to L^2(\Gamma)$. It can be expressed as a matrix with respect to a basis set of $L^2(\Gamma)$. Let the set $\{\psi_i(\boldsymbol{q},\boldsymbol{p})\}$

5.1 Zwanzig's projection formalism

be a orthonormal basis set of $L^2(\Gamma)$. The dynamical variable A can be written as a linear combination of the basis vectors (which are L^2-functions, $\psi_i : \Gamma \to \mathbb{R}$),

$$A(\boldsymbol{q}, \boldsymbol{p}, t) = \sum_i a_i(t) \psi_i(\boldsymbol{q}, \boldsymbol{p}). \tag{5.4}$$

The scalar product acts as a projection to the basis vectors. Therefore, the coefficients of the expansion of A can be obtained as $a_i = \langle \psi_i | A \rangle$. The Liouville equation can be written componentwise as

$$\frac{\partial}{\partial t} a_k(t) = \sum_i L_{ki} a_i(t). \tag{5.5}$$

The matrix elements of the Liouville operator with respect to the basis $\{\psi_i\}$ are obtained as

$$L_{ki} = \langle \psi_k | \mathbf{L} \psi_i \rangle. \tag{5.6}$$

Note that the Hilbert space $L^2(\Gamma)$ is of infinite dimension, irrespective of how many dimensions span Γ. Therefore, the matrices must be compact to ensure the convergence of the infinite sums in the above calculations. Here, we are not dealing with the mathematical difficulties of this sort.

When describing physical systems with many degrees of freedom one often assumes that not all details of the system are equally important. A lot of the dynamical information can be deemed unimportant or irrelevant for the features of interest. We seek a physical description based on those quantities considered relevant and neglect the irrelevant details. The properties of the system, which we would like to understand, are certainly relevant, but it may turn out that further quantities can not be ignored as they crucially determine the time evolution. There is neither a general rule, how to find a set of relevant quantities or coordinates in a given physical system, nor can it be taken for granted that a useful reduced description exists. But the reduced description has proved advantageous in many situations and can sometimes considerably increase the efficiency of the physical description.

If we chose a set of relevant coordinates, the time evolution can be obtained by a projection of Eq. (5.1) to the subspace spanned by the relevant coordinates. The effort to express the time evolution of dynamical quantities in terms of basis functions in L^2 allows projection operators to be written in a simple form. In what follows, we assume a two-dimensional situation: one relevant coordinate and one irrelevant coordinate. The Liouville equation, Eq. (5.1) reads in the two-dimensional representation,

$$\frac{\partial}{\partial t}\begin{pmatrix}a_1\\a_2\end{pmatrix}=\begin{pmatrix}L_{11}&L_{12}\\L_{21}&L_{22}\end{pmatrix}\begin{pmatrix}a_1\\a_2\end{pmatrix}. \tag{5.7}$$

The solution of the second component can be written as

$$a_2(t) = e^{L_{22}t}a_2(0) + \int_0^t e^{L_{22}(t-s)}L_{21}a_1(s)\mathrm{d}s. \tag{5.8}$$

The solution of $a_2(t)$ can be substituted into the first component of the Liouville equation, leading to

$$\frac{\partial}{\partial t}a_1(t) = L_{11}a_1(t) + L_{12}\int_0^t e^{L_{22}(t-s)}L_{21}a_1(s)\mathrm{d}s + L_{12}e^{L_{22}t}a_2(0). \tag{5.9}$$

The irrelevant coordinate enters only with its initial condition in the last term, the rest of the equation depends only on a_1, the relevant coordinate. Eq. (5.9) can be read as an equation depending on a_1 with a perturbation represented by the last term, which is also referred to as noise term, $\xi(t) = L_{12}e^{L_{22}t}a_2(0)$. With the definitions $\mu(t) = -e^{L_{22}t}L_{21}$, $\Omega = L_{11}$, and $\gamma = L_{12}$ the equation for the relevant variable $v(t) = a_1(t)$ reads

$$\frac{\partial}{\partial t}v = \Omega v - \gamma \int_0^t \mu(t-s)v(s)\mathrm{d}s + \xi(t). \tag{5.10}$$

This is the *generalized Langevin equation* (GLE). The time evolution of the relevant coordinate v is given by an equation that has some similarity with the Langevin equation, Eq. (2.11). The Langevin equation is the limiting case of Eq. (5.10) for $\Omega = 0$, $\mu(t) = \delta(t)$ and assuming ξ to be white noise. In contrast to the classical Langevin equation, the GLE is not a differential equation but an integro-differential equation, which is not local in time, *i.e.*, the derivative of v at time t^* depends on the values of v in the past, $t < t^*$. Therefore, the description of the dynamics of v in the reduced subspace is non-Markovian! The integral with the memory function $\mu(t)$ represents correlations in the system. This is the consequence of the projection. In other words: reducing the complexity of a Markovian system by a projection to a subspace gives rise to non-Markovian dynamics and a perturbation term. Eq. (5.10) gives also a justification of the noise term in the free Langevin equation, Eq. (2.11): the description in a relevant subspace gives rise to stochasticity.

The general case, where the projection leads to a multidimensional subspace, fading out several dimensions, can be treated by a similar approach. In Eq. (5.9), the a_i are two parts of the vector A, containing the relevant coordinates for $i = 1$ and the irrelevant coordinates for $i = 2$. Reading Eq. (5.9) as a matrix equation, the memory function is a consequence of the off-diagonal blocks in the Liouville operator, L_{21} and L_{12}, which couple the relevant with the irrelevant coordinates. For details, see [69].

The remarkable result of this section is that whenever we ignore parts of a Hamiltonian system, we introduce a stochastic element in the dynamics; and we are confronted with non-Markovian dynamics as a consequence of incomplete dynamical information. The analysis of the MD simulations in Chap. 4 was based on the degrees of freedom of the simulated biomolecule (β-hairpin or $(GS)_nW$ peptide). However, in the explicit water simulation the molecule is just a subsystem, the dynamics of which depend on the dynamics of the surrounding water. Therefore, we have to take account of the possibility of memory effects and non-Markovian behavior.

The analysis of the different water models used in the MD simulation of a β-hairpin molecule indicate that the memory effects due to he water dynamics are not significant on the time scale 1 ps, see Sec. 4.3.

If external memory effects are to be taken into account in MD simulation, random forces with correlated noise ("colored noise") can be used, as was recently suggested by Ceriotti et al. [148].

5.2 Chain dynamics

MD simulations are based on a refined, empiric, classical model of the microscopic dynamics. An interpretation of the MD trajectories and their features is impeded by the complexity of the model used. Therefore, simplified models can help to identify the mechanism that gives rise to a specific dynamical effect. At best, this allows to check which of the "ingredients" of the model account for which sort of dynamical properties.

A common model is the *Rouse chain*, developed in the context of polymer physics [149, 150]. The Rouse chain consists of N beads, each of mass m. Let $\mathbf{z}_i$ be the position of bead i. The beads are connected by Hookean springs with the angular frequency $\tilde{\omega}$, such that they built a linear chain without branching. In the overdamped approximation, *i.e.*, ignoring inertial effects, the Hamilton function of the chain reads

$$H = \sum_{i=1}^{N} \frac{1}{2} m \tilde{\omega}^2 \left[(\mathbf{z}_{i-1} - \mathbf{z}_i)^2 + (\mathbf{z}_i - \mathbf{z}_{i+1})^2 \right]. \tag{5.11}$$

As the three Cartesian coordinates of each bead decouple, the Rouse chain is treated here as a one-dimensional problem, replacing the vectors $\mathbf{z}_n$ by z_n. The chain dynamics can be described in terms of the normal modes, ζ_n, which are derived in Appendix C. The eigenfrequencies obtained from NMA are given as

$$\omega_n = \frac{\tilde{\omega} \pi n}{N - 1} \quad \text{with } n = 0, 1, ..., N - 1. \tag{5.12}$$

The Hamilton function reads in the basis of the normal modes

$$H = \frac{1}{2} m \sum_{n=0}^{N-1} \omega_n^2 \zeta_n^2(t). \tag{5.13}$$

Note that the normal modes collapse with the PC, as the beads have identical mass m and all interaction potentials are harmonic.

To include thermal fluctuations, the chain is coupled to a Langevin heat bath with friction, γ, and a frequency specific random force, ξ_n, analogous to Eq. (2.11). In the normal mode coordinates the equations of motion read

$$m \gamma \frac{\mathrm{d}}{\mathrm{d}t} \zeta_n = -m \omega^2 \zeta_n + \xi_n, \tag{5.14}$$

where the inertial terms are ignored as in the Hamilton function (overdamped approximation). The noise ξ_n is assumed to be white noise [Eqs. (2.12) and (2.13)].

5.2 Chain dynamics

Therefore, we can treat the system as ergodic and do not need to specify the average procedure in what follows in this section.

As the $\zeta_n(t)$ contribute quadratically to the Hamilton function, the equipartition theorem can be applied. Hence, the ACF of the normal modes is given as

$$\langle \zeta_n(t+\tau)\zeta_k(\tau)\rangle = \frac{k_B T}{m\omega_n^2} e^{-\omega_n^2 t/\gamma} \delta_{nk}. \qquad (5.15)$$

Using Eq. (2.27), the MSD of the normal modes follows from the ACF as

$$\langle [\zeta_n(t+\tau) - \zeta_n(\tau)]^2 \rangle = 2\frac{k_B T}{m\omega_n^2}\left(1 - e^{-\omega_n^2 t/\gamma}\right). \qquad (5.16)$$

Hence, the MSD of the normal modes/PC modes exhibits a linear time dependence for short times and saturates for long times. The typical saturation time of the slowest mode,

$$\tau_R = \frac{\gamma(N-1)^2}{\pi^2 \tilde{\omega}^2}, \qquad (5.17)$$

is the so-called *Rouse* time, which dominates the long time behavior of the model. Finite time averaging, $\langle \cdot \rangle_{\tau,T}$, needs to exceed the Rouse time, *i.e.* $T \gg \tau_R$, in order to justify the application of the ergodic hypothesis.

From Eq. (5.16) it follows that the MSD of the normal modes is not subdiffusive in the Rouse model. However, other generalized coordinates can exhibit subdiffusive behavior. Consider the distance between two beads, $\Delta(t) = z_i(t) - z_j(t)$. The beads i and j are assumed to be neither neighbored nor close to the chain ends. If $t \ll \tau_R$, the ACF of $\Delta(t)$ is given as

$$\langle \Delta(t+\tau)\Delta(\tau)\rangle_\tau = t^{-1/2}. \qquad (5.18)$$

A detailed derivation is given in Appendix C. The power-law decay of the ACF leads immediately to a subdiffusive MSD according to Eq. (2.27). The subdiffusive distance fluctuations are a consequence of a superposition of the exponential normal modes. Since the long-time dynamics are given by the slowest mode, the power-law ACF, Eq. (5.18), breaks down at $\approx \tau_R$, with an exponential decay for $t > \tau_R$ [151].

The above calculations for the Rouse chain can be extended to include hydrodynamic effects [152] or long-time memory effects using a GLE approach

[153, 154]. The model can also be generalized to other bead spring geometries, which include branching, loops and fractal clusters [151, 155, 156]. It can be demonstrated that this allows power laws to occur in the distance ACF with various exponents [155].

The structure of the backbone suggests the application of chain models to the dynamics of peptides [153, 157]. GLE based models have been employed to model long-time dynamics of biomolecules [153, 154]. Cross links and more sophisticated coupling geometries between the beads have been designed for the description of protein fluctuations [151, 155].

Distance fluctuations, as exemplified by the above derivation, are of particular interest: fluorescence quenching by a tryptophan residue can be used to measure the distance fluctuations in single molecule experiments [27, 51, 52, 157]. The observation of a flavin reductase protein complex revealed subdiffusive dynamics on the 10^{-4} to 10^0 s time scale [27, 51, 52]. However, an estimation of the Rouse time is in the order of a few nanoseconds, *i.e.*, at least four orders of magnitude below the observed subdiffusive regime [158].

In the case of the biomolecules presented in Chap. 4, typical normal mode fluctuations are in the range of some squared nanometers, *i.e.* $\langle \Delta x^2 \rangle \approx 1\,\text{nm}^2$. Approximately, the corresponding frequencies are obtained as $\tilde{\omega}^2 \approx 3k_B T/m \langle \Delta x^2 \rangle$, which are in the range of $\tilde{\omega} \approx 1\,\text{ps}^{-1}$. A typical friction value of a heavy atom at the surface of an biomolecule is $\gamma \approx 50\,\text{ps}^{-1}$ [159]. Eq. (5.17) allows the Rouse time of the longest biomolecule in Chap. 4 ($N = 16$, the number of residues) to be estimated as $\tau_R \approx 76\,\text{ps}$. In contrast, the subdiffusion found in the $(GS)_n W$ peptides and the β-hairpin extends to 10 ns. Therefore, the Rouse chain model – irrespective of potential subdiffusive dynamics on short time scales – cannot provide an understanding of the subdiffusive dynamics as found in the MD simulations presented in Sec. 4.2.

5.3 The Continuous Time Random Walk

Trap models

Non-exponential relaxation patterns, as found in the dynamics of proteins and peptides, are a common property of glassy materials. There are other characteristics shared by glasses and proteins, such as non-Arrhenius temperature dependence [29, 30] and the enhancement of fluctuations above the glass transition [31]. The similarity of biomolecules and glasses is considered as a consequence of various, nearly isoenergetic conformational substates that both, proteins and glasses, can assume [20, 23, 30]. The energy landscape of glasses, and the "glassy" dynamics that follows from it, were successfully described as a hopping between energy traps on a fully connected lattice [160, 161]. Energy traps are understood as local minima of the energy landscape [24]. Kramer's escape theory [69] allows to ascribe a typical escape time to a potential energy minimum at a given temperature. The typical time of escaping can be identified with an effective energy depth of the minimum. Then, the distribution of escape times, or, equivalently, the distribution of effective trap depths determines the dynamics of the system [160, 162]. The essential properties of this sort of trap models are equivalently reproduced by the continuous time random walk (CTRW) model [9].

Introduction of the CTRW model

Pearson's random walk, as discussed in Sec. 2.4, has a fixed jump length, Δx and a fixed time step, Δt. Here, such random walks with fixed, discrete time step are referred to as *classical* random walks. The diffusion limit corresponds to the behavior on length scales $\gg \Delta x$ and time scales $\gg \Delta t$. In mathematical terms, this is expressed by the limit $(\Delta x, \Delta t) \to 0$, in which $D = \Delta x^2/2\Delta t$ is constant. The CTRW generalizes the classical random walk, such that both variables, jump length and time step are random variables. The jump length, here denoted as x, is characterized by the jump length distribution (JLD), $\varphi(x)$. Throughout this thesis it is assumed, that the JLD is symmetric around $x = 0$ and the mean value of the JLD equals zero. The time between two successive jumps is referred to as *waiting time*, t, and it is taken from the distribution $w(t)$.

Assume the JLD has the variance $\bar{x}^2$ and the waiting time distribution (WTD) has the mean value $\bar{t}$. Then, a derivation analogous to Sec. 2.4 can be applied to derive the diffusion equation, Eq. (2.33), in the limit $\bar{x}^2 \to 0$ and $\bar{t} \to 0$ with $D = \bar{x}^2/2\bar{t}$ constant. Therefore, if both, the variance of the JLD and the mean value of the WTD, are finite, classical diffusion occurs, *i.e.*, Eq. (2.33) is valid and the MSD exhibits a linear time dependence,

$$\left\langle \Delta x^2(t) \right\rangle = 2Dt, \tag{5.19}$$

as in Eq. (2.35).

The situation is different, if the distributions do not satisfy the above conditions. If the JLD has a diverging variance, the CTRW is found to be superdiffusive, *i.e.*, the MSD exhibits a time dependence $\sim t^\beta$, where $\beta > 1$ [8]. A subdiffusive MSD occurs, if the WTD has an infinite mean value.

The present thesis is focused on subdiffusion. Therefore, we confine the discussion to the following scenario, as laid out in [8]. We study a one-dimensional[1] CTRW with the JLD, $\varphi(x)$ and the WTD, $w(t)$. The JLD is assumed to have mean value zero, to be symmetric, and to have a finite variance $\bar{x}^2$. Let the WTD have a power-law tail, *i.e.*,

$$w(t) \sim \left(\frac{t}{\tau_0}\right)^{-1-\alpha}, \tag{5.20}$$

with $0 < \alpha < 1$ referred to as the *WTD exponent*. The time τ_0 defines the time unit and is not a relaxation time. As a consequence of the power-law decay of $w(t)$, the distribution has a diverging mean value. The results in this section are independent of the detailed analytical form of $w(t)$ at short times. A first jump from the position $x = 0$ is assumed to occur at $t = 0$. Note that this are two independent initial conditions: the initial condition for the position is $W_0(x) = \delta(x)$, *i.e.*, being at $x = 0$ at $t = 0$, whereas the initial condition for the waiting times is having the first jump at $t = 0$. Illustrations of one-dimensional CTRWs with various WTD exponents are given in Fig. 5.1.

---

[1] Multidimensional CTRWs can be decomposed into a set of one-dimensional CTRWs, if the walk is statistically isotropic.

Application of CTRW

The CTRW has been developed in the context of charge currents in semiconductors [9, 90, 163] or, more generally, for transport in amorphous, disordered materials [164]. There is a plethora of applications in very different fields, see [8, 74, 90] for a review. Trapping models have become a powerful model applied to a variety of different processes, such as diffusion through actin networks [84, 165], diffusion crossing a membrane [166], or through porous media [85, 86]. There is a wide variety of cases where, rather than unbounded diffusion, distinct boundaries exist and can have critical effects on diffusive dynamics [167]. Examples of these are the subdiffusive dynamics of macromolecules in the cell nucleus [168], the cytoplasm, which has been shown to emerge from crowding [78, 80–82, 169], in cell and plasma membranes [170–173], and the subdiffusion of lipid granules, which is influenced by the presence of entangled actin networks [79, 83, 84]. The consequences of reactive boundaries [174–176] and diffusion through different kinds of narrow pores and tubes [177–182] have also been examined.

In the context of internal dynamics of biomolecules, CTRW has been suggested as a possible mechanism causing the dynamics to be subdiffusive [52, 183], based on the analysis of time-averaged quantities, such as the ACF. However, care must be taken when applying the CTRW model to time averages of measured time series, because the CTRW is an intrinsically non-ergodic process. Recently, the question as to how time-averaged CTRW quantities behave has attracted theoretical attention [165, 184–188]. In this section, the time averages of CTRW are examined. The time dependence of the time-averaged MSD for unbounded and for bounded diffusion are among the results obtained in the present thesis [140, 147].

Simulation of CTRW

The theoretical results in this sections are compared to simulations of a CTRW with WTD exponent α. The simulations were performed as follows. The jump lengths are taken from a Gaussian distribution with variance one, implemented as a Box-Muller algorithm [107]. All CTRWs were started at $x = 0$ with an initial jump at $t = 0$. A random variable t_w with the distribution given in Eq. (5.20) is obtained from the uniformly distributed $r \in (0,1)$ via the transformation $t_w = r^{-1/\alpha} - 1$. With this transformation, the WTD has the following analytical form

$$w(t_w) = \frac{\alpha}{(1+t_w)^{1+\alpha}}. \tag{5.21}$$

Here, we set $\tau_0 = 1$. Uniformly distributed random numbers, $r \in (0,1)$ were generated with the long-period random number generator of L'Ecuyer with Bays-Durham shuffle and added safeguards [107].

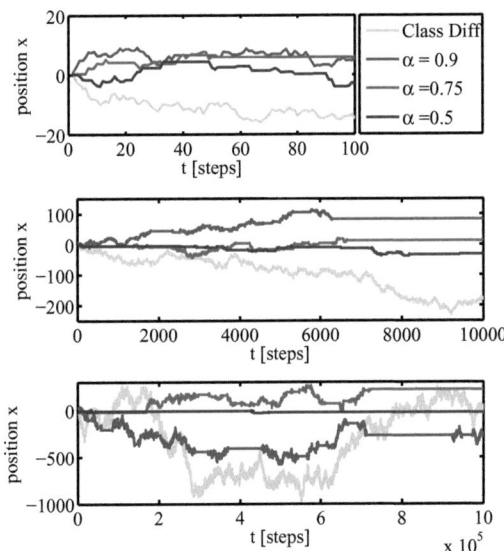

Figure 5.1: (Color online) **CTRW – individual time series** - Time series of the positions, $x(t)$, of individual CTRWs are displayed. The colors refer to various WTD exponents, $\alpha = 0.5$ (blue), $\alpha = 0.75$ (red), and $\alpha = 0.9$ (green). In a classical random walk with the same $\bar{x}^2$. The three figures illustrate three different time scales, $T = 10^2, 10^4, 10^6$ (top down). The CTRWs exhibit on all time scales a similar pattern of waiting times.

5.3.1 Ensemble averages and CTRW

In Sec. 2.4, the diffusion equation is derived from the classical random walk. Instead of a microscopic description of the random walk, the diffusion equation determines the time evolution of the probability distribution, $W(x,t)$. For a large number, N, of similar particles, the dynamics of which is given by $W(x,t)$, the probability gives rise to a particle density, $\rho(x,t) = NW(x,t)$. Note that the probability $W(x,t)$ corresponds to an ensemble average. In contrast to the individual particle, the density is a macroscopic quantity. With this perspective, the diffusion equation is the macroscopic description of a large number of microscopic, classical random walkers. The question arises as to whether a similar equation exists which describes the time evolution of the probability of a CTRW. It has been demonstrated that such an equation can be derived from the CTRW, *e.g.*, by means of Fourier-Laplace transforms [8], which is demonstrated in Appendix D together with some mathematical details.

In the case of CTRW, the diffusion limit reads $(\bar{x}, \tau_0) \to 0$, such that the generalized diffusion constant, defined as $K_\alpha = \bar{x}^2/[2\Gamma(1-\alpha)\tau_0^\alpha]$, has a finite, non-zero value. The fractional diffusion equation (FDE) is [6]

$$\frac{\partial}{\partial t} W(x,t) = {}_0\mathcal{D}_t^{1-\alpha} K_\alpha \frac{\partial^2}{\partial x^2} W(x,t), \qquad (5.22)$$

where the Riemann-Liouville operator is used

$$ {}_0\mathcal{D}_t^{1-\alpha} \phi(t) = \frac{1}{\Gamma(\alpha)} \frac{\partial}{\partial t} \int_0^t \frac{\phi(t')}{(t-t')^{1-\alpha}} \mathrm{d}t'. \qquad (5.23)$$

$\Gamma(\alpha)$ is the Gamma function, see Appendix D.1.

The derivation of Eq. (5.22) involves an averaging procedure over the ensemble. The elements of the ensemble are all possible realizations of CTRW respecting the initial condition, in particular, all members of the ensemble have a first jump at $t = 0$. Obviously, there is no other point in time, at which the probability of observing a jump equals one. Therefore, the initial condition breaks the time-shift invariance of the CTRW process. This symmetry breaking is referred to when saying that the CTRW in the ensemble average undergoes *aging*. Eq. (5.22) is an integro-differential equation, *i.e.*, it is not a local equation in time: the time derivative at t is influenced by the values of $W(x, t')$ with $t' \leqslant t$. This accounts for memory effects and makes the process described by Eq. (5.22) non-Markovian. As the memory effects do not decay fast, *i.e.*, the decay follows a power law instead of an exponential decay, the process is non-ergodic.

As in the case of the diffusion equation, Fourier decomposition allows to give an analytical solution of Eq. (5.22) in terms of the Mittag-Leffler function[2] (MLF)

$$W(x,t) = \int_0^\infty a(k) E_\alpha(-K_\alpha k^2 t^\alpha) e^{-ikx} dk. \tag{5.25}$$

The Fourier coefficients $a(k)$ depend on the initial condition of the position and the boundary conditions. If $W_0(x) = \delta(x)$, the coefficient function is $a(k) = (2\pi)^{-1}$. The position part of Eq. (5.25) is identical to the classical expression in Eq. (2.34). The different behavior is due to the Mittag-Leffler decay which replaces the exponential relaxation of the diffusion equation [Eq. (2.33)]. The MLF exhibits an asymptotic power-law dependence at long times, $E_\alpha(-K_\alpha k^2 t^\alpha) \sim t^{-\alpha}$, which is considerably slower than an exponential decay. This causes long-lasting memory effects. In particular, the power-law relaxation pattern does not possess a typical time scale, a consequence of the WTD from Eq. (5.20). Therefore, there is no time scale on which correlations can be ignored. Hence, time-averages over an interval $[t_s, t_s + T]$ depend on t_s, irrespective of the length of T. The ergodic hypothesis is not applicable[3], the CTRW exhibits *weak ergodicity breaking* [184–186].

---

[2] The Mittag-Leffler function (named after Gösta M.-L. (1846-1927), Swedish mathematician) is defined as

$$E_\alpha(z) = \sum_{n=0}^{\infty} \frac{z^n}{\Gamma(1+n\alpha)}. \tag{5.24}$$

For details see Appendix D.

[3] As mentioned for the classical diffusion equation, there is no meaningful stationary solution for the position probability distribution in the absence of a potential or a boundary. Therefore, care must be taken with the ergodic hypothesis when dealing with free, unbounded diffusion. The same argument applies to Eq. (5.22) representing the diffusion limit of the free, unbounded CTRW. When energy potentials are present, the FDE can be generalized to a fractional Fokker-Planck equation, which contains Eq. (5.22) as a limiting case [7]. With appropriate potentials, a stationary solution exists for both, the classical random walk and the subdiffusive CTRW. However, even with a stationary solution, the memory effects embodied by Eq. (5.25) are fundamentally in conflict with the ergodic hypothesis. See Subsec. 5.3.3.

5.3 The Continuous Time Random Walk

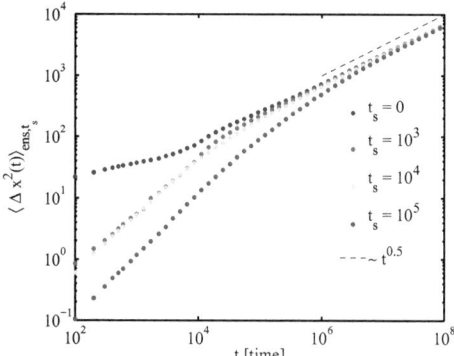

Figure 5.2: (Color online) **Free, unbounded CTRW – ensemble-averaged MSD** - Ensemble-averaged MSD of an CTRW with WTD exponent $\alpha = 0.5$. The MSD exhibits a power-law time dependence, $\langle \Delta x^2(t) \rangle_{ens,0} \sim t^\alpha$, as predicted by Eq. (5.26). However, the shifted MSD, $\langle \Delta x^2(t) \rangle_{ens,t_s}$, depends on t_s, i.e., it is not invariant with time shift. The ensemble average is performed over 1 000 individual CTRWs.

Eq. (5.22) leads to a subdiffusive MSD

$$\langle \Delta x^2(t) \rangle_{ens,0} = \int_{-\infty}^{\infty} x^2 W(x,t) \mathrm{d}x = \frac{2K_\alpha}{\Gamma(1+\alpha)} t^\alpha, \qquad (5.26)$$

which can be derived by means of Fourier-Laplace transformation [8]. The notation $\langle \cdot \rangle_{ens,0}$ indicates averages over the ensemble with the initial condition of a first jump at $t = 0$ during the time interval $[0, t]$. Therefore, Eq. (5.26) is the ensemble-averaged MSD from the origin, $x = 0$, during the time interval $[0, t]$. The ensemble-averaged MSD is not invariant with time shift. The ensemble-averaged MSD during the time interval $[t_s, t + t_s]$ with $t_s > 0$ of a CTRW with initial jump at $t = 0$, denoted by $\langle \Delta x^2(t) \rangle_{ens,t_s}$, is different from $\langle \Delta x^2(t) \rangle_{ens,0}$. In Fig. 5.2 simulation results with several values of t_s illustrate the non-stationarity of the process.

Due to the breaking of the time shift symmetry of the CTRW, the process depends on the time of the first jump. Loosely speaking, the process 'looks statistically different' at different times; it is not stationary.

Assume we observe an CTRW process, starting the observation at a time

$t_s > 0$. The time t_s is almost always between two jumps. Therefore, a period t_1 elapses until we observe the first jump. The distribution of the initial waiting times t_1 is denoted as $w_1(t_1, t_s)$. The time shift invariance makes the initial WTD, $w_1(t_1, t_s)$ depend on t_s. After the initial waiting time has elapsed, a first jump occurs at $t_s + t_1$. If a new time coordinate is introduced by $\tilde{t} = t - t_s - t_1$, the first jump takes place at $\tilde{t} = 0$. In this time coordinate the usual CTRW theory can be applied for times $\tilde{t} > 0$. Hence, the initial WTD is sufficient to characterize the situation, in which the observation starts at t_s. A CTRW with $t_s > 0$ is referred to as aging CTRW (ACTRW) [189].

5.3.2 Time averages and CTRW

The CTRW is a non-ergodic process. As the mean of the WTD in Eq. (5.20) diverges, there is no typical relaxation time in CTRW. As a consequence, time averages of CTRW quantities are, in general, different from the ensemble average of the same quantities. Although many types of experimental measurements provide ensemble averages, certain techniques provide time series, such as, for example, single-particle tracking, single molecule spectroscopy or, in particular, the MD simulations presented in Chap. 3. Time averages are then required to extract statistically significant properties from the data. Therefore, the question has arisen as to how the time-averaged properties of CTRW processes behave [140, 165, 184–188].

In the following we focus on the relative probability of being at position x at a time $t > t_s$, provided the walker was at position x_s at time $t_s > 0$. This probability can be expressed as [190, 191]

$$W(x, t; x_s, t_s) = \sum_{n=0}^{\infty} W_n(x; x_s) \chi_n(t; t_s), \qquad (5.27)$$

in which $W_n(x; x_s)$ is the probability of reaching x from x_s in exactly n jumps, and $\chi_n(t; t_s)$ is the probability of making exactly n jumps in the time interval $[t_s, t]$. The variable n is referred to as *operational time*.

Eq. (5.27) decomposes the CTRW into terms depending on its two stochastic ingredients, the WTD, $w(t)$ and the JLD, $\varphi(x)$. Since $\varphi(x)$ has a finite variance,

5.3 The Continuous Time Random Walk

$\bar{x}^2$, the probability $W_n(x; x_s)$ is the same as in a random walk, in which the walker jumps with a fixed frequency but the jump length is a random variable with distribution $\varphi(x)$. An approximation of $W_n(x; x_s)$ can be obtained as the solution of the diffusion equation, Eq. (2.34), with $D = \bar{x}^2/2$. The probability $W_n(x; x_s)$ represents an average over all possible series of jump lengths from the JLD $\varphi(x)$. The probability $\chi_n(t; t_s)$ reflects the waiting time as a random variable, that is, χ_n is a probability with respect to an average over all possible series of waiting times from $w(t)$.

An outstanding property of the CTRW model is the fact that the time averages of CTRW quantities are random variables, as will be illustrated with the following argument. In this paragraph, let the position coordinate, x, of a CTRW be restricted to the integer values $x = 1, 2, ..., K$. The random walker jumps from x only to $x + 1$ or $x - 1$, with a probability q_x or $1 - q_x$, respectively. At $x = 1$ and $x = K$ the walker jumps always to the right and left, respectively. The waiting times between two jumps are a random variable characterized by the WTD Eq. (5.20). The probability of being at x after n jumps is denoted by $P_n(x)$. The process is Markovian and in the operational time, it is a classical random walk. Therefore, it can be described with a Master equation [187, 192]. Hence, an equilibrium is reached for large n, i.e. $\lim_{n \to \infty} P_n(x) = p_x$. The operational time spent at x is denoted as N_x, the total number of jumps is $N = \sum_x N_x$, and, for large N, $p_x = N_x/N$. Let T_x be the time spent at position x, the total observation period is $T = \sum_x T_x$, in the regular time coordinate. An observable $\phi(x)$, being a function of x, has the time average over the period $[0, T]$

$$\langle \phi \rangle_{\tau,T} = \sum_x w_x \phi(x), \tag{5.28}$$

in which $w_x = T_x/T$ is the average sojourn time at the position x. It can be demonstrated that the T_x are random variables with a common distribution function that follows from Lévy statistics [187]. As a consequence, the time average in Eq. (5.28) is a random variable [184, 185]. An example is illustrated in Fig. 5.3 A. In the present thesis, the distributions of time-averaged quantities are not examined. Instead, we focus on mean values of time-averaged quantities. Using mean values of time-averaged CTRW quantities allows the application of

Eq. (5.27), which involves ensemble averaging. Hence, in what follows the combined ensemble- and time-averaged MSD in the interval $[0, T]$ is derived.

We now return to the free, unbounded CTRW with continuous position space. Eq. (5.27) is used to characterize the CTRW when it is observed at time t_s. The decomposition of the CTRW into its stochastic ingredients is exploited to obtain the ensemble-averaged MSD between time t_s and the time $t_s + t$, which is given as

$$\left\langle [x(t_s + t) - x(t_s)]^2 \right\rangle_{ens,0} = \int_{-\infty}^{\infty} (x - x_s)^2 W(x, t + t_s; x_s, t_s) \mathrm{d}x. \quad (5.29)$$

With Eq. (5.27) this can be expressed as

$$\left\langle [x(t_s + t) - x(t_s)]^2 \right\rangle_{ens,0} = \sum_{n=0}^{\infty} \left\langle \Delta x^2(n) \right\rangle \chi_n(t; t_s), \quad (5.30)$$

where the MSD of the free, unbounded, classical random walk in operational time with JLD $\varphi(x)$ is used [see Eq. (2.35)],

$$\left\langle \Delta x^2(n) \right\rangle = \int_{-\infty}^{\infty} W_n(x; x_s) \mathrm{d}x = 2Dn. \quad (5.31)$$

This is a first step towards the time-averaged MSD, which is obtained by performing an average over $t_s \in [0, T]$ on both sides of Eq. (5.30). Thus the time-averaged MSD over the observation period $[0, T]$ follows as

$$\left\langle \Delta x^2(t) \right\rangle_t = \frac{1}{T - t} \int_0^{T-t} \left\langle [x(t + \tau) - x(\tau)]^2 \right\rangle_{ens,0} \mathrm{d}\tau, \quad (5.32)$$

where the combined ensemble-time average is defined as

$$\langle \cdot \rangle_t := \left\langle \langle \cdot \rangle_{ens,0} \right\rangle_{t_s, T}. \quad (5.33)$$

Note that the time average and the ensemble average in Eq. (5.33) can be swapped. The t_s-dependence in Eq. (5.27) is due to χ_n. The t_s-dependence of χ_n is due to the initial WTD, w_1, in Eq. (5.38). Therefore, a time average in the interval $[0, T]$ can be performed by substituting $w_1(t, t_s)$ using the time-averaged initial WTD

$$\bar{w}_1(t) = \langle w_1(t, \tau) \rangle_{\tau, T} = \frac{1}{T} \int_0^T w_1(t, \tau) \mathrm{d}\tau. \quad (5.34)$$

5.3 The Continuous Time Random Walk

The form of the initial WTD $w_1(t, t_s)$ is derived in [189]. It has the long t asymptotic behavior

$$w_1(t, t_s) \sim \frac{\sin \pi\alpha}{\pi} \frac{1}{t + t_s} \frac{t_s^\alpha}{t^\alpha}. \tag{5.35}$$

The time-averaged initial WTD is obtained by integration over Eq. (5.35) in the limit $t \ll T$ [147]

$$\bar{w}_1(t) \approx \frac{\kappa_\alpha}{T^{1-\alpha}} t^{-\alpha}, \tag{5.36}$$

where the constant $\kappa_\alpha = \sin(\pi\alpha)/\pi\alpha$ is introduced.

The probability of making exactly n jumps during a time period of length t is the joint probability of making $n - 1$ jumps in a shorter time $t - \tau$ and of finding a waiting time τ, integrated over all possible values of τ. More precisely, the probability of making $n \geqslant 2$ jumps in the time interval $[t_s, t_s + t]$ is given by

$$\chi_n(t + t_s; t_s) = \int_0^t \chi_{n-1}(\tau + t_s; t_s) w(t - \tau) \mathrm{d}\tau. \tag{5.37}$$

The probability of making no or one jump in $[t_s, t + t_s]$ can be expressed as

$$\chi_1(t + t_s; t_s) = \int_0^t w_1(\tau, t_s) w(t - \tau) \mathrm{d}\tau, \text{ and} \tag{5.38}$$

$$\chi_0(t + t_s; t_s) = 1 - \int_0^t \mathrm{d}\tau w_1(\tau, t_s) \mathrm{d}\tau. \tag{5.39}$$

Eqs. (5.37) and (5.38) allow an iterative expression to be established, in which χ_n is expressed as an $n - 1$-fold convolution integral.

In order to evaluate Eqs. (5.37) – (5.39), the Laplace transform $t \to u$ is applied. In the Laplace representation, the $n-1$-fold convolution for χ_n corresponds to a product

$$\chi_n(u; t_s) = \frac{1 - w(u)}{u} [w(u)]^{n-1} w_1(u, t_s) \qquad \text{for } n \geqslant 1. \tag{5.40}$$

For $n = 0$, $\chi_0(u, t_s) = [1 - w_1(u, t_s)]/u$. With Eqs. (5.40) and (5.31) the ensemble-averaged MSD between t_s and $t_s + t$ in Eq. (5.30) is expressed in the Laplace representation as

$$\mathcal{L}\left[\langle [x(t_s + t) - x(t_s)]^2 \rangle_{ens,0}\right](u) = 2Dw_1 \frac{1-w}{u} \sum_{n=0}^{\infty} n w^{n-1} \tag{5.41}$$

$$= 2D \frac{w_1}{u[1-w]}, \tag{5.42}$$

where the geometric series is used and the functional arguments, u and t_s, are skipped. As the t_s dependence is entirely due to w_1, the time-averaged initial WTD $\bar{w}_1$ is substituted in Eq. (5.40), *i.e.*,

$$\bar{\chi}_n(u) = \langle \chi_n(u;t_s) \rangle_{t_s,T} = \frac{1-w(u)}{u}[w(u)]^{n-1}\bar{w}_1(u) \qquad \text{for } n \geqslant 1. \qquad (5.43)$$

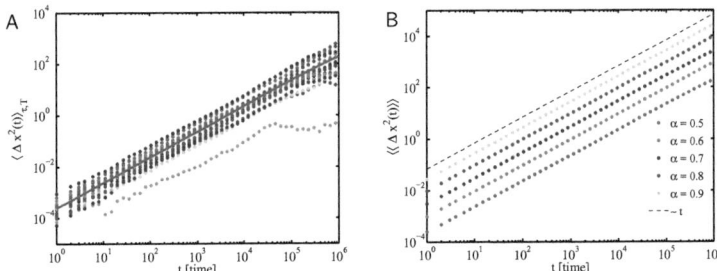

Figure 5.3: *(Color online)* **Free, unbounded CTRW – time-averaged MSD.** - **A**: Individual time-averaged MSD, $\langle \Delta x^2(t) \rangle_{\tau,T}$, of an CTRW with WTD exponent $\alpha = 0.5$. The continuous red line is the average over 100 MSDs, $\langle \Delta x^2(t) \rangle_{t}$. The individual MSDs scatter around $\langle \Delta x^2(t) \rangle_{t}$, as the time-averaged CTRW quantities are random variables. **B**: Time-averaged MSDs of CTRWs with various WTD exponents α (see legend). The MSD, $\langle \Delta x^2(t) \rangle_{t}$, is obtained of individual trajectories as in **A**. An additional averaging is performed over 100 individual MSDs. Independent of α, all MSDs exhibit a linear time dependence. No subdiffusive behavior is present for the time-averaged MSD of a free, unbounded CTRW, as predicted by Eq. (5.47). Note that a different notation for double average is used in the axis label, $\langle\langle \cdot \rangle\rangle$ instead of $\langle \cdot \rangle_t$.

In the Laplace transform, the long t asymptotic behavior corresponds to small u behavior in the Laplace space. In particular, the value $u = 0$ corresponds to the integral over the entire t-range. The Laplace transform of the WTD $w(t)$ with power-law decay as in Eq. (5.20) reads

$$w(u) \sim 1 - \Gamma(1-\alpha)u^\alpha \qquad \text{with } u \gtrsim 0. \qquad (5.44)$$

For the time-averaged initial WTD in Eq. (5.36), the Laplace transform has the form

$$\bar{w}_1(u) \sim \frac{(uT)^{\alpha-1}}{\Gamma(1+\alpha)} \qquad \text{with } u \gtrsim 0. \qquad (5.45)$$

5.3 The Continuous Time Random Walk

With the Laplace transforms of w and $\bar{w}_1$, the right side of Eq. (5.42), upon averaging over t_s, has the following form

$$\left\langle \Delta x^2(u) \right\rangle_{\mathrm{t}} = 2D \frac{K_\alpha}{T^{1-\alpha}} u^{-2}. \tag{5.46}$$

The time-averaged, ensemble-averaged MSD in the interval $[0,T]$ of the CTRW with WTD exponent α follows upon inverse Laplace transformation of Eq. (5.46) as

$$\left\langle \Delta x^2(t) \right\rangle_{\mathrm{t}} = 2D \frac{K_\alpha}{T^{1-\alpha}} t. \tag{5.47}$$

In contrast to the ensemble-averaged MSD of Eq. (5.26), the time-averaged MSD is *not* subdiffusive. The linear time dependence of the time-averaged MSD is illustrated in Fig. 5.3 for various WTD exponents α. The MSD, $\left\langle \Delta x^2(t) \right\rangle_{\mathrm{t}}$ depends on the length of the observation of the process.

An alternative way to calculate the average $\left\langle \Delta x^2(t) \right\rangle_{\mathrm{t}}$ is to exploit the fact that after the initial waiting time, when the first jump occurs, the ensemble average can be obtained as the solution of the FDE, *i.e.* from Eq. (5.25).

$$\left\langle \Delta x^2(t) \right\rangle_{\mathrm{t}} = \int_0^t \bar{w}_1(\tau) \left\langle \Delta x^2(t-\tau) \right\rangle_{ens,0} \mathrm{d}\tau. \tag{5.48}$$

The convolution integral in Eq. (5.48) can be evaluated in the Laplace representation as a product. The MSD $\left\langle \Delta x^2(t) \right\rangle_{ens,0}$ is obtained from the solution of the fractional diffusion equation, Eq. (5.25).

5.3.3 Confined CTRW

The time-averaged MSD of the free, unbounded CTRW was found to not exhibit subdiffusive behavior [Eq. (5.47)]. However, the possible applications suggest that, rather than unbounded diffusion, finite volume effects may have a critical influence on the properties of a CTRW. In this subsection, the *confined* CTRW is examined, in which the walker is restricted to the interval $x \in [0,L]$ by reflecting boundaries.

The ensemble-averaged, time-averaged MSD, $\left\langle \Delta x^2(t) \right\rangle_{\mathrm{t}}$, can be derived analogously to the last subsection, if the unbounded MSD in operational time, $\left\langle \Delta x^2(n) \right\rangle =$

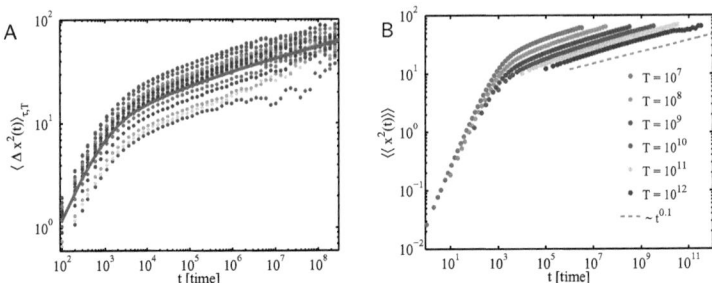

Figure 5.4: (Color online) **Confined CTRW – time-averaged MSD** - A: The dotted lines are time-averaged MSDs, $\langle \Delta x^2(t) \rangle_{\tau,T}$, of individual trajectories with WTD exponent, $\alpha = 0.9$, simulation length, $T = 10^9$, and $L = 20$. The continuous red line is the ensemble-average over 1000 of the individual MSDs, $\langle \Delta x^2(t) \rangle_t$. The individual MSDs exhibit a common underlying pattern: a linear time dependence for short times and cross over to a subdiffusive time dependence, $\sim t^{1-\alpha}$, for long times. However, the individual $\langle \Delta x^2(t) \rangle_{\tau,T}$ do not coincide, as time averages of CTRWs are random variables. B: Ensemble-averaged, time-averaged MSD, $\langle \Delta x^2(t) \rangle_t$ for various simulation lengths T (WTD exponent $\alpha = 0.9$ and boundary width $L = 20$). In order that each time series should contain the same number of points (10^7), the time resolution of the longer simulations was reduced. Therefore, the short time behavior of the MSDs of larger T is absent. The MSDs display a dependency on the simulation length and decrease with increasing T. $\langle x^2(t) \rangle_t$ is bounded by the value $L^2/6$. For longer T, the time-averages MSD is shifted to smaller values, but does not reach a constant plateau. Note that a different notation for the double average is used in the axis label, $\langle\langle \cdot \rangle\rangle$ instead of $\langle \cdot \rangle_t$.

$2Dn$, is replaced with the corresponding finite volume MSD[4] in Eq. (5.42). The finite volume MSD is given in Eq. (A.8). Then, Eq. (5.42) reads

$$\langle \Delta x^2(u) \rangle_t = \frac{L^2}{6} \bar{w}_1 \frac{1-w}{uw} \sum_{n=0}^{\infty} w^{n-1} \left[1 - g^n\right] \qquad (5.49)$$

$$= \frac{L^2}{6} \frac{\bar{w}_1}{uw} \frac{1-g}{1-gw}, \qquad (5.50)$$

where $g = \exp[-\pi^2 D/L^2]$.

---

[4] Here, the approximation Eq. (A.8) is used rather than Eq. (A.5), for simplicity. It turns out that the approximation does not affect the results inappropriately. However, the mathematical argument can equally be applied to the exact solution Eq. (A.5).

5.3 The Continuous Time Random Walk

Eq. (5.50) can be evaluated for two limiting cases: the $u \to 0$ behavior determines the long t asymptotic form of $\langle \Delta x^2(t) \rangle_t$, whereas $u \lesssim 1$ corresponds to the short time limit, $t \gtrsim 1$. For small $u \ll 1$, Eq. (5.50) reads

$$\langle \Delta x^2(u) \rangle_t \approx \frac{L^2}{6} \frac{1}{T^{1-\alpha}} \frac{u^{\alpha-2}}{\Gamma(\alpha+1)}. \tag{5.51}$$

Hence, the long t behavior of the time-averaged MSD is given as

$$\langle \Delta x^2(t) \rangle_t \approx \frac{L^2}{6} \frac{\kappa_\alpha}{1-\alpha} \left(\frac{t}{T}\right)^{1-\alpha}. \tag{5.52}$$

The short-time behavior is obtained from the Laplace representation in the range $u \lesssim 1$

$$\langle \Delta x^2(u) \rangle_t \approx \frac{L^2}{6} \frac{\kappa_\alpha}{T^{1-\alpha}} \frac{1-g}{g} u^{-2}. \tag{5.53}$$

With $(1-g)/g \approx 12D/L^2$, the time-averaged MSD at short times has the following t dependence

$$\langle \Delta x^2(t) \rangle_t \approx 2D \frac{\kappa_\alpha}{T^{1-\alpha}} t. \tag{5.54}$$

The time-averaged MSD of the confined CTRW exhibits a two-phasic behavior. In contrast to the free, unbounded CTRW, the confined CTRW has a subdiffusive time-averaged MSD at long t. The subdiffusion found has an exponent $1-\alpha$. For short times, the result of the free, unbounded CTRW, Eq. (5.47), is retrieved, as for short times the boundary does not affect the MSD of the random walker. The two phases of $\langle \Delta x^2(t) \rangle_t$ are separated from each other by a critical time, which is given as

$$t_c = \left(\frac{L^2}{12D(1-\alpha)}\right)^{\frac{1}{\alpha}}. \tag{5.55}$$

The critical time, t_c, does not depend on the length of the simulation, T. Up to a constant factor, the time t_c can be understood as the time, after which the ensemble-averaged MSD of the free, unbounded CTRW equals the plateau value. Therefore, the critical time cannot depend on the observation length, T. The subdiffusive behavior, $\langle x^2(t) \rangle_t \sim t^{1-\alpha}$, found in Eq. (5.52) contrasts with the $\sim t^\alpha$ behavior reported by He et al. [165], which arose from the fact that the time

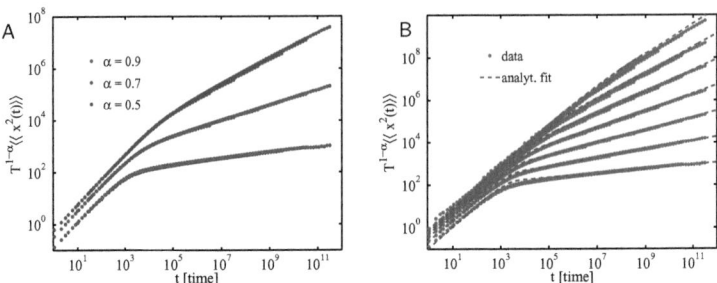

Figure 5.5: (Color online) **Confined CTRW – Simulation length scaling and analytical fit** - A: Ensemble-averaged, time-averaged MSD, $T^{1-\alpha} \langle \Delta x^2(t) \rangle_t$ of various $T = 10^8, 10^9, ..., 10^{14}$ for each $\alpha = 0.5$, 0.7 and 0.9 ($L = 20$). The MSD is multiplied with $T^{1-\alpha}$ to remove the T-dependence. The coincidence of the data for different simulation lengths, T, illustrates the correct scaling behavior of Eq. (5.56). B: The T-scaled MSD, $T^{1-\alpha} \langle \Delta x^2(t) \rangle_t$, as in A is compared to the analytical expression in Eq. (5.56) for $\alpha = 0.3, 0.4, ..., 0.9$ (top down). The good fit to the simulation data of the analytical curve demonstrates the validity of the approximations made in the derivation of Eq. (5.56). It also illustrates that the critical time, t_c, as given in Eq. (5.55), does not depend on the simulation length, T.

window, in which $\sim t^\alpha$ was fitted to the MSD in [165], was confined to times close to $\sim t_c \approx 27\,000$.

The following interpolation between short and long t domains can be performed

$$\langle \Delta x^2(t) \rangle_t = \frac{L^2}{6} \frac{\kappa_\alpha t^{1-\alpha}}{(1-\alpha)T^{1-\alpha}} \left[1 - \exp\left(-\frac{1-\alpha}{L^2} 12 D t^\alpha \right) \right]. \quad (5.56)$$

5.3.4 Application of CTRW to internal biomolecular dynamics

The CTRW model has been suggested as a possible mechanism for the subdiffusive MSD found in the internal dynamics of proteins [52, 183]. Subdiffusive behavior is found in single molecule fluorescence spectroscopy [27, 28, 52] and spin echo neutron scattering [54], experimental techniques providing time-averaged ACFs.

5.3 The Continuous Time Random Walk

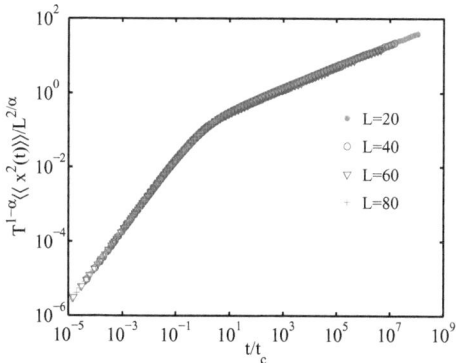

Figure 5.6: *(Color online)* **Confined CTRW, scaling with volume** - Ensemble-averaged, time-averaged MSD, $T^{1-\alpha} \langle \Delta x^2(t) \rangle_t / L^{2/\alpha}$ of various $L = 20, 40, 60,$ and 80 and various $T = 10^8, ..., 10^{13}$ with $\alpha = 0.7$. The figure illustrates the L-scaling of $\langle \Delta x^2(t) \rangle_t$ as given in Eq. (5.56). It also demonstrates the L-dependence of the critical time, t_c, Eq. (5.55).

Thus, the experiments revealed subdiffusion in the time-averaged ACF. Likewise, MD simulations indicating subdiffusive dynamics provide time-averaged MSDs [53, 140]. However, the subdiffusive MSD found in a free, unbounded CTRW is obtained after ensemble averaging, cf. Eq. (5.26), whereas the time-averaged MSD is not subdiffusive! Therefore, the free, unbounded CTRW cannot explain the subdiffusion in the above mentioned cases. The confined CTRW exhibits a subdiffusive, time-averaged MSD for times $t > t_c$. Therefore, the confined CTRW, in contrast to the free, unbounded CTRW, could explain the MD results (and the subdiffusion of the time-averaged ACF found in experiment).

The fact that the time-averaged MSD obtained from MD simulation reaches a saturation plateau, whereas the time-averaged MSD of the confined CTRW does not, does not necessarily disqualify the confined CTRW as a model for MD simulations. The CTRW can be modified such that the power-law WTD of Eq. (5.20) holds up to a time scale t_{max} (*i.e.*, for $t < t_{max}$ Eq. (5.20) is valid), but beyond which (*i.e.* for $t > t_{max}$) the WTD decays faster than t^{-2}. Equilibrium would be reached on a time scale $t > t_{max}$. In this case, for $t < t_{max}$ the process behaves as a CTRW, whereas for $t > t_{max}$ classical diffusion occurs. Subdiffusion is found on time scales $t_c < t < t_{max}$, (*i.e.*, times shorter than t_{max} but long enough for the system to explore the accessible volume). For $t < t_c$ a linear time dependence occurs.

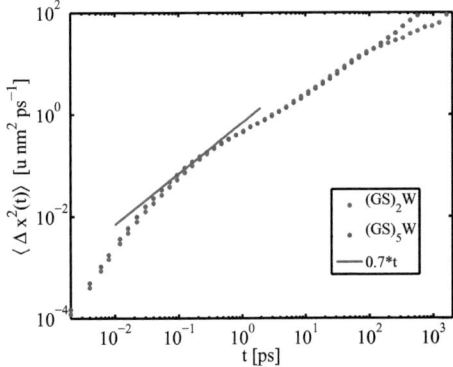

Figure 5.7: *(Color online)* **MSD of $(GS)_2W$ and $(GS)_5W$ at short time scales.** - The MSD grows ballistically at the time scale of the MD integration step $\Delta t = 2\,\text{fs}$. At around $t = 0.3\,\text{ps}$, it crosses over from a ballistic to a subdiffusive time dependence. The tangent to the curve, $2Dt$ with $D = 0.35\,\text{u nm}^2\,\text{ps}^{-1}$, illustrated as full line colored in magenta, allows the upper bound of the diffusion constant to be estimated. Note that the estimation, independent of the system under investigation, leads to the same diffusion constant; the diffusion in the configuration space is isotropic as long as the diffusion is not affected by the presence of the potential energy landscape.

When assuming the confined CTRW model for the MSD as found in the MD simulations of Chap. 4, the diffusion constant cannot be obtained: none of the simulations has a time resolution which allows the linear time dependence of the MSD to be observed, but the subdiffusive part does not depend on D. To estimate the diffusion constant, an MD simulation was performed with an increased time resolution; the coordinates were recorded with 2 fs time resolution, corresponding to the MD integration time step (instead of the time resolution of 1 ps used in the $(GS)_nW$ simulations and of 2 ps used for the β-hairpin). The total length of this simulation is 2 ns. The MSD of the simulation is illustrated in Fig. 5.7. For the shortest time scales the MSD grows ballistically, as expected from inertial effects. On the time scale of inertial effects, the CTRW model does not apply; no inertial effects are included in CTRW. The MSD in Fig. 5.7 does not exhibit a linear time dependence, *i.e.*, the MSD crosses over from the ballistic to the subdiffusive regime. However, we estimate an upper bound of the diffusion constant, D, choosing the smallest D for which $\langle \Delta x^2(t) \rangle \leqslant 2Dt$ in Fig. 5.7. The approximation of $D \lesssim 0.35 \, \mathrm{u\,nm^2\,ps^{-1}}$ is based on the assumption that at the time scale, on which the MSD crosses over from ballistic to subdiffusive, *i.e.* around $\approx 0.3\,\mathrm{ps}$, the MSD exhibits, in principal, a linear time dependence. The diffusion constant reflects the diffusion in the configuration space for time scales, on which the diffusion is not affected by the form of the energy landscape, *i.e.*, time scales on which the configurational diffusion can be considered as free and unbounded. At these short time scales, the diffusion is assumed to be isotropic due to the equipartition theorem, which states that all degrees of freedom can be treated in a similar way using the harmonic approximation around a local minimum of the energy landscape. Therefore, the diffusion constant is expected to be independent of the system under investigation. This independence is confirmed by Fig. 5.7, as for both, $(GS)_2 W$ and $(GS)_5 W$ a very similar diffusion is observed on short time scales.

We assumed classical diffusion for estimation of the diffusion constant. Why not use Eq. (5.54) for the fit? The CTRW can only be understood as a rough sketch of the dynamics. Obviously, the configuration, *i.e.* the configuration space coordinate, never is strictly at rest but always fluctuates. Waiting times can only be introduced as periods during which the overall displacement in the configuration space does not exceed a specified value. During such periods the systems appears to be trapped. Hence, the WTD can be imagined as a consequence of local minima of the energy landscape acting as kinetic traps. However, at short times, the molecule can be assumed to perform unhindered, isotropic diffusion, even if it is just inside a single local minimum of the energy landscape. Again, the diffusion on short time scales is independent of the system under investigation.

We calculate the critical times, t_c, for the individual PCs. The time t_c of the individual PCs is compared to the time, t_p, on which the MSD of the PC mode reaches a constant value. Fig. 5.8 illustrates the critical times of the individual PCs of the $(GS)_n W$ as crosses (+), while the saturation times, t_p, are given as

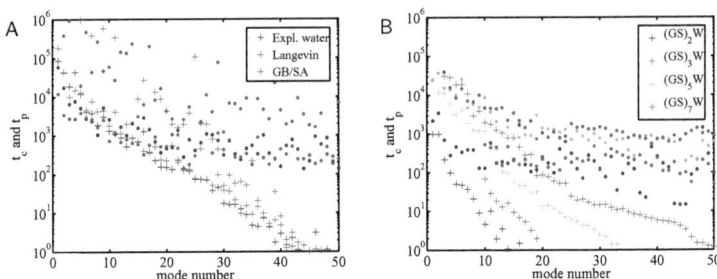

Figure 5.8: *(Color online)* **Critical time of confined CTRW of $(GS)_nW$ peptides.** - The crosses, $+$, illustrate the critical time as function of the PCs and the dots represent the times, t_p, on which the PCs saturates to the plateau of saturation. **A:** The different colors refer to the different water models in the simulation of the β-hairpin, see legend. **B:** The different colors refer to the $(GS)_nW$ of different $n = 2, 3, 5,$ and 7, see legend. The dots illustrate the times after which the MSD, $\langle \Delta x^2(t) \rangle$, reaches the constant plateau. For the lowest PCs the critical time, t_c, lies in the order of the saturation time t_p. Even for intermediate PCs the critical time $t_c > 1$. Therefore, the subdiffusion seen in the lowest PCs ranging from 1 ps to t_p cannot be explained by the confined CTRW model.

dots. For the lowest PCs, t_p and t_c are in the same order of magnitude. The confined CTRW model predicts subdiffusive behavior only in the range $t_c < t < t_p$. Therefore, the confined CTRW model cannot account for the subdiffusion seen in the low PCs of the β-hairpin and the $(GS)_nW$ peptides.

5.4 Diffusion on networks

In the theory of complex systems, the application of network models has become a common approach [193, 194]. Networks connect the theory of Markov chains to transport processes in fractal geometries and in amorphous or disordered media. As subdiffusion is a typical feature of random walks on fractal geometries, networks are a promising approach for finding the mechanisms that give rise to subdiffusive behavior.

5.4.1 From the energy landscape to transition networks

Assume a conservative, mechanical system. The configuration of the system is given by the vector x. The interaction potential between different parts of the system give rise to a potential energy, which is a function of x. In addition to the dynamics determined by the potential energy, the system is interacting with its environment, modeled as a velocity-dependent friction and a random force, given as stationary white noise. This scenario corresponds to the classical Langevin equation, Eq. (2.11), in multiple dimensions with an arbitrary underlying potential. We assume that the fluctuation-dissipation theorem holds. The friction and the white noise represent the coupling to a heat bath, and the system is assumed to be in thermal equilibrium. For each configuration of the internal coordinates, x, the system has a certain potential energy due to the interaction potential between the systems constituents. The white noise causes the system to populate the potential energy landscape according to Boltzmann statistics. The kinetics of the system arises entirely from the potential energy landscape. If the dynamics of the system is recorded with a time resolution Δt, for each $t = n\Delta t$ a vector containing all internal coordinates, $x(t) = x_n$, characterizes the time evolution of the system. If the accessible volume of the configuration space is finite, the trajectory is long enough, and Δt sufficiently small, then the trajectory allows the potential energy landscape to be reconstructed.

An approximate representation of the potential energy landscape can be obtained as follows [195]. First, the configuration space is partitioned into discrete states. Then, each possible configuration is attributed to one of the discrete states, $e.g.$, to the most similar discrete state in the configuration space. The similarity follows from the metric used in the configuration space. In doing so, the trajectory, $(x_1, x_2, ..., x_T)$ is mapped onto a sequence of discrete states, $(\kappa_1, \kappa_2, ..., \kappa_T)$. If the states are seen as the vertices of a network, the transitions between the states define the edges of the network. The network represents the possible transitions between the discrete states on a specified time scale. If a weight factor is assigned to each of the edges, this factor can be chosen such that it characterizes the relative frequency of the corresponding transition. Such a network is referred to as a transition network.

The transition network represents the features of the potential energy landscape with arbitrary precision, if the number of discrete states is adequate; in the limit of an infinite number of states, the potential energy landscape is retained from the transition matrix. Using the network representation of the potential energy landscape as input of a Markov model, the complete dynamical behavior can be reproduced.

Here, we try to characterize the energy landscape of biomolecules in terms of transition networks. An MD trajectory of a molecule with explicit water can be seen as an example of the above scenario. However, if the dynamics are projected to the internal coordinates of the molecule, the projection may lead to memory effects as demonstrated in Sec. 5.1. The projected equilibrium probability distribution follows from the *free* energy landscape. In contrast to the *potential* energy landscape, the free energy landscape is, in general, insufficient to determine the kinetic behavior. From the discussion in Sec. 4.3, it follows that the memory effects due to the water dynamics are small on the time scale of 1 ps. Therefore, a Markov model may still capture the dynamics of the projected trajectory above 1 ps.

5.4.2 Networks as Markov chain models

Let us start with a set of states, *i.e.*, regions of the configuration space, $\mathcal{N} = \{n_1, n_2, \ldots \ldots, n_N\}$, and a discrete trajectory $(\kappa_1, \kappa_2, \ldots, \kappa_T)$, obtained from the full configuration space dynamics, $\boldsymbol{x}(t) = \boldsymbol{x}(n\Delta t) = \boldsymbol{x}_n$, as described in the last paragraph. The counting matrix, $\mathbf{Z}(\vartheta)$, contains as components z_{ij} the number of transitions from state j to state i on the time scale $\vartheta = \nu \Delta t$,

$$z_{ij} = \sum_{n=1}^{T} \delta[\kappa_n - j]\delta[\kappa_{n+\nu} - i]. \tag{5.57}$$

In general, for a finite trajectory, the counting matrix is not symmetric.

A thermodynamics system tends to an equilibrium state at long times. The equilibrium is characterized by a stationary phase space density. As said above, the dynamics of a Hamiltonian system in contact with a heat bath can be modeled with a multidimensional Langevin equation and the full interaction potential

5.4 Diffusion on networks

$V(\boldsymbol{x})$,
$$\frac{d^2}{dt^2}\boldsymbol{x} - \Gamma\frac{d}{dt}\boldsymbol{x} + \boldsymbol{\xi} - \nabla_x V(\boldsymbol{x}) = 0. \tag{5.58}$$
or a sufficiently strong friction, the overdamped case, the equation reduces to
$$\Gamma\frac{d}{dt}\boldsymbol{x} + \nabla_x V(\boldsymbol{x}) = \boldsymbol{\xi}. \tag{5.59}$$
Eq. (5.59) allows the velocity coordinate to be decoupled from the position coordinate. Therefore, we are justified to assume an equilibrium state in which the probability of being in a volume $d^{3N}x$ at position $\boldsymbol{x}$ is proportional to
$$\exp\left(-\frac{V(\boldsymbol{x})}{k_B T}\right) d^{3N} x. \tag{5.60}$$
The stationary probability distribution of the discretized trajectory is denoted by the vector $\boldsymbol{p}^s$, with the components p_i^s being the equilibrium probability of being in state i. The equilibrium probability density in the configuration space translates to the equilibrium probability of the discretized trajectory as
$$p_i^s = \mathcal{Z}_{config}^{-1} \int_{V_i} \exp\left(-\frac{V(\boldsymbol{x})}{k_B T}\right) d^{3N} x, \tag{5.61}$$
with the configurational partition function
$$\mathcal{Z}_{config} = \int \exp\left(-\frac{V(\boldsymbol{x})}{k_B T}\right) d^{3N} x. \tag{5.62}$$

In equilibrium, the number of transitions from i to any other state equals the number of transitions from any state to i. Furthermore, if the number of transitions from i to j were unequal to those of the transitions from j to i, $z_{ij} \neq z_{ji}$, then the equilibrium dynamics in configuration space would contain loops, *i.e.*, closed paths exhibiting a preferred rotational direction. The presence of such vortices with preferred rotational sense in the configuration space would be in conflict with the second law, as such eddies would constitute, in principle, a *perpetuum mobile*, a perpetual motion machine of the second kind. Therefore, in the limit of $T \to \infty$, $\mathbf{Z}(\vartheta)$ is symmetric, $z_{ij} = z_{ji}$. The term *detailed balance* refers to the latter equality.

The transition matrix, $\mathbf{S}(\vartheta)$, contains as components, s_{ij}, the relative probability of being in state i at time $t + \vartheta$ given that at the time t the system was in state j. It follows $\sum_j s_{ij} = 1$ for all i. An approximate transition matrix, $\tilde{\mathbf{S}}(\vartheta)$, can be obtained from the counting matrix as $\tilde{s}_{ij} = z_{ij}/\sum_k z_{kj}$. It follows that the transition matrix can be approximated on the basis of a finite discrete trajectory $\kappa(t) = \kappa(n\Delta t) = \kappa_n$ as

$$s_{ij} \approx \tilde{s}_{ij} = \frac{\sum_{n=1}^{T} \delta[\kappa_n - j]\delta[\kappa_{n+\nu} - i]}{\sum_{m=1}^{T} \delta[\kappa_m - j]}, \qquad (5.63)$$

where $\vartheta = \nu\Delta t$. The transition matrix characterizes the kinetics on the time scale ϑ. The matrix $\mathbf{S}(\vartheta)$ does not contain information about the dynamics on time scales below ϑ. If we assume that the probability of finding state i at time $t^* + \vartheta$ depends only on the state at time t^*, but not on the states at the times $t < t^*$, then the transition matrix furnishes complete dynamical information with respect to the set of discrete states chosen. If the history of the system does not influence the present state, *i.e.*, the events at times $t < t^*$ do not affect the relative probability, the system is Markovian.

In Sec. 5.1 it is demonstrated that the Markov property of any system may be lost upon projection to a relevant, low dimensional subspace. Therefore, we cannot expect the dynamics of a solvated peptide or protein to be Markovian in the internal coordinates of the biomolecule, as we neglect the influence of the solvent dynamics. However, the Markov property may be a beneficial approximation on time scales on which memory effects have sufficiently decayed. It is demonstrated in Chap. 4 that the Markov property is approximately satisfied on the time scale $\lesssim 1\,\mathrm{ps}$ for the systems discussed in the present thesis. Note, however, that the projection to subspaces of fewer dimensions gives rise to further memory effects, which may be relevant on time scales above $1\,\mathrm{ps}$.

If the Markov property is a valid approximation on and above a time scale t_m, the dynamics for $t > \vartheta$ can be represented by the transition matrix, $\mathbf{S}(\vartheta)$, where $\vartheta \geqslant t_m$.

5.4.3 Subdiffusion in fractal networks

Often, diffusion processes occur in disordered materials, *e.g.*, the diffusion in porous media [85, 86] or the charge transport in amorphous semiconductors [9]. Fractal lattices proved a beneficial model of disordered media and paved the way

5.4 Diffusion on networks

to a rigorous mathematical analysis of the dynamics in fractal geometries. The concept of self-similarity – introduced by Benoît Mandelbrot – is the key property of fractal objects which allows to apply scaling arguments. The self-similarity over different length scales gives rise to a diffusion behavior similar over different time scales. Therefore, power-law behavior is a typical feature of fractal behavior. In what follows, we briefly review the concept of fractality. A more detailed discussion of the topic can be found in the review articles [162, 196].

Geometrical fractality

Geometrical objects can be classified by their dimension: a line is a one-dimensional object, a plane is two-dimensional and a cube is three-dimensional. There are several concepts in mathematics to introduce the quantity *dimension*. The topological dimension of a geometrical object is the minimal number of parameters needed to characterize uniquely a single point of the object[5]. For example, the position on the earth's surface can be parametrized by two parameters, *e.g.*, geographical longitude and latitude. Hence, the surface of the earth is a two-dimensional object. The topological dimension has always an integer value.

In 1918, the German mathematician Felix Hausdorff introduced a dimension definition based on measure theory. The Hausdorff dimension coincides for most objects like cubes, planes or curves with the topological dimension, but can have non-integer values for certain, irregular objects. A simplified approach to the Hausdorff dimension is the following argument. A geometrical object in a d-dimensional space shall be covered by d-dimensional spheres of radius R. Let $N(R)$ be the minimal number of spheres needed to cover the object at a given R. The Hausdorff or *fractal* dimension is defined as

$$d_f = -\lim_{R \to 0} \frac{\log N(R)}{\log R}. \tag{5.64}$$

Fractals are defined as objects for which the fractal dimension is unequal to the topological dimension, $d_f \neq d_t$. Less rigorously, the fractal dimension can be

---

[5] The topological dimension, also covering or Lebesgue dimension, of a subset Ξ of a topological space M is strictly defined as the minimal $d_t \in \mathbb{N}$ such that every open cover of Ξ has a refinement, in which every point $x \in \Xi$ is included in $d_t + 1$ or fewer open sets.

understood in the following way. Usually the volume of a (massive) object in a d-dimensional space scales with the length as $V \sim L^d$. In contrast, if the fractal can be imagined as a massive body, it has a volume scaling as

$$V \sim L^{d_f}. \qquad (5.65)$$

For example, the volume excluded by a fractal in a cube of side length L is typically $V \sim L^{d_f}$.

The Sierpiński triangles and the Sierpiński gasket – introduced by the Polish mathematician Waclaw Sierpiński in 1915 – are used here as examples of a regular fractal and as toy models for the diffusion in disordered media. The first order Sierpiński triangle, Γ_1, is an equilateral triangle of side length one Fig. 5.9 A. The second order Sierpiński triangle, Γ_2, is constructed by the combination of three first order Sierpiński triangles, the lengths of which are scaled by $1/2$, as illustrated in Fig. 5.9. The construction scheme is repeated each time the order is increased by one, cf. Fig. 5.9. The limit $\Gamma_\infty = \lim_{n \to \infty} \Gamma_n$ yields the Sierpiński gasket.

Figure 5.9: **Construction of the Sierpiński gasket** - *A: First order Sierpiński triangle, Γ_1. B: Second order Sierpiński triangle, Γ_2. C: Third order Sierpiński triangle, Γ_3. D: Sierpiński triangle of order eight, Γ_8. The Sierpiński gasket is obtained as $\Gamma_\infty = \lim_{n \to \infty} \Gamma_n$.*

The Sierpiński gasket has the topological dimension one, $d_t = 1$. The Sierpiński gasket contains three copies of itself, each down-scaled by a factor of one half. Therefore, the fractal dimension can be derived from $2^{d_f} = 3$ to be $d_f = \log 3 / \log 2 \approx 1.5850$. For the Sierpiński gasket the Hausdorff dimension exceeds the topological dimension, $d_f > d_t$, *i.e.*, the Sierpiński gasket is a fractal. Note that the dimension of the embedding space, here the two-dimensional plane, exceeds the fractal dimension.

5.4 Diffusion on networks

The argument used to derive the fractal dimension of the Sierpiński gasket refers to its self-similarity. By construction, Γ_∞ contains identical copies of itself on arbitrary small length scales. Therefore, the Sierpiński gasket has no typical length scale, it is a scale free object – apart from the finite size of Γ_∞. The regular and deterministic way to construct the Sierpiński gasket allows renormalization techniques to be applied.

The non-integer Hausdorff dimension motivated the term *fractal* coined by Mandelbrot. Although all objects with non-integer Hausdorff dimension are fractals, not all fractals have a non-integer fractal dimension d_f. An important example is the random walk. Assume a random walker in a two-dimensional plane makes a jump every $\Delta t = 1\,\text{s}$. The jump length may be a random variable with a Gaussian distribution of variance $\sigma^2 = 1\,\text{m}^2$ and all directions are equally likely. Due to the fact that the sum of independent, Gaussian random variables is again a Gaussian random variable, the displacements of the walker on the time scale one second are statistically the same as those on the time scale two seconds, provided the variance of the two-second-displacement is scaled by a factor $\sqrt{2}$. In the diffusion limit, *i.e.*, $\Delta t \to 0$ and $\sigma^2 \to 0$ with $D = \sigma^2/2\Delta t$ constant, the fractal dimension of the trajectory of the random walker is given by $2^{d_f/2} = 2$, *i.e.*, $d_f = 2$. Since the topological dimension of the random walkers trajectory is $d_t = 1$, the trajectory, seen as a geometrical object in the embedding space, is a fractal. Note that the fractal dimension equals two, irrespective of the dimension of the embedding space. If the random walk is embedded to a two-dimensional space, *i.e.* the random walk in the plane, the process is room-filling. The fractal dimension is reflected by the fact that the MSD of the classical, unbounded random walk in space of arbitrary dimension is linear in time [cf. Eqs. (2.18) and (2.35)]. Guigas *et al.* have conjectured that the subdiffusive motion of macromolecules in the cytoplasm increases the fractal dimension of the trajectory, d_f, such that the trajectories are room filling in the three-dimensional space [82].

Self-similarity is a surprisingly abundant property in the real world, some of the most impressive examples being the Romanesco broccoli cabbage, fern plants, the shape of clouds and the surface structure of rocks. However, the strict self-similarity on all length scales, as found in mathematics, cannot be observed in nature. Rather, a statistical self-similarity, as in the case of the random walker occurs in nature. Furthermore, self-similarity usually is found in a range of length-scales. In that sense, the trajectories of physical Brownian particles are fractal above the length scale defined by the mean free path.

So far, the fractal quantity of geometrical objects has been discussed. Now, we proceed to dynamical processes on fractal objects.

Dynamical fractality

De Gennes raised the question as to how a random walk on a percolation cluster, the path of an "ant in the labyrinth", would look like [197, 198]. As fractals turned out to be suitable model systems for irregular, amorphous media, and being similar to percolation clusters [199], the random walk on fractal geometries became an active field of research in the 1980's. A review on diffusion in amorphous media can be found in [164].

Random walks on fractals can be modeled as random walks on transition networks. Let a transition network have the nodes $\mathcal{N} = \{n_1, n_2, ..., n_N\}$ and the edges $\mathcal{E} \subset \mathcal{N} \times \mathcal{N}$. A random walker may jump in unit time from one node, n_i, to another node, n_j, if n_i is connected to n_j with an edge, i.e., $(n_i, n_j) \in \mathcal{E}$. If the edges are of equal weight, the walker, being at n_i, chooses with uniform probability one of the neighbores of n_i, i.e., one of the nodes which share an edge with n_i. If the edges are weighted by a transition matrix, $\mathbf{S}$, then the probability of jumping to node n_j from n_i is given as s_{ji}. Note that the transition matrix is defined such, that $\sum_i s_{ji} = 1$ for all j. The random walk on the network forms a Markov chain.

Let the network be embedded to a d-dimensional vector space, V, each node, n_i, having a position $\bm{v}_i \in V$. The jump of the random walker on an embedded network from node n_i to node n_j corresponds to a displacement $\bm{u} = \bm{v}_j - \bm{v}_i$. Therefore, the random walk on the network can be mapped to a series of displacements in the embedding space, $\bm{u}_1, \bm{u}_2, ..., \bm{u}_T$. The displacement from the starting position after t jumps is $\sum_{i=1}^{t} \bm{u}_i$. With the vector space metric, the MSD can be calculated as

$$\langle \Delta x^2(t) \rangle = \left\langle \left(\sum_{i=1}^{t} \bm{u}_i \right)^2 \right\rangle = \sum_{i=1}^{t} \langle \bm{u}_i^2 \rangle + 2 \sum_{i=1}^{t} \sum_{j>i}^{t} \langle \bm{u}_i \bm{u}_j \rangle . \qquad (5.66)$$

5.4 Diffusion on networks

For a large network with nodes homogeneously distributed in the embedding space, the mean square of the single displacement is $\langle \boldsymbol{u}^2 \rangle$ and the MSD reads

$$\langle \Delta x^2(t) \rangle = \langle \boldsymbol{u}^2 \rangle t + 2 \sum_{i=1}^{t} \sum_{j>i}^{t} \langle \boldsymbol{u}_i \boldsymbol{u}_j \rangle. \tag{5.67}$$

For networks in which the edges can be considered as uncorrelated, one has $\langle \Delta x^2(t) \rangle = \langle \boldsymbol{u}^2 \rangle t$. The linear time dependence is found in the case of regular lattices and corresponds to the MSD of the classical random walk, Eq. (2.35).

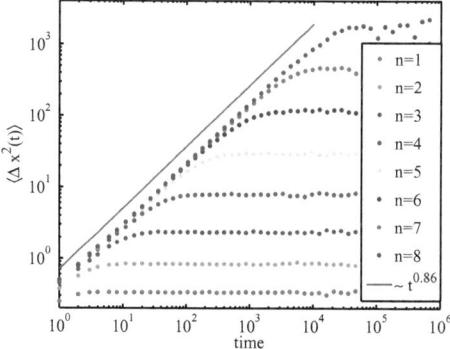

Figure 5.10: (Color online) **MSD of Sierpiński triangles** - The MSD of a random walk on the Sierpiński triangle Γ_n is illustrated for various $n = 1, 2, ..., 8$ (see legend). The full red line illustrates the power law with the exponent $\alpha = 0.86$, as predicted from the renormalization argument.

However, fractal networks usually do not have uncorrelated displacements, i.e. $\langle \boldsymbol{u}_i \boldsymbol{u}_j \rangle \neq 0$. Rather than a linear time dependence, the MSD of random walkers on fractal networks is found to be subdiffusive, i.e.

$$\langle \Delta x^2(t) \rangle \sim t^{2/d_w}, \tag{5.68}$$

where $d_w > 2$ is the *diffusion dimension* of the fractal network. Note that the MSD exponent $\alpha = 2/d_w$ conveys the same information as the diffusion dimension, d_w. For all finite fractal networks, i.e., networks with a finite number of nodes, the MSD is bounded. Therefore, Eq. (5.64) holds below the time scale on which the MSD saturates to its maximal value. The subdiffusion found in fractal networks is due to the geometry of the network and is not a violation of the Markov property. Fractal networks can exhibit subdiffusion in equilibrium and as an ergodic phenomenon. An example is illustrated in Fig. 5.10, in which the MSD of random walkers on the Sierpiński triangle, Γ_n, for various n is illustrated.

The diffusion dimension of the Sierpiński gasket can be derived with a renormalization argument. In the construction of the Sierpiński gasket going from Γ_n to Γ_{n+1} corresponds to the replacement of each triangle by the second order Sierpiński triangle, Γ_2, the side length of which is twice as much as for the triangle, Fig. 5.11. We compare the typical transit time in the triangle, τ, to the typical transit time of the second order Sierpiński triangle, τ' [196].

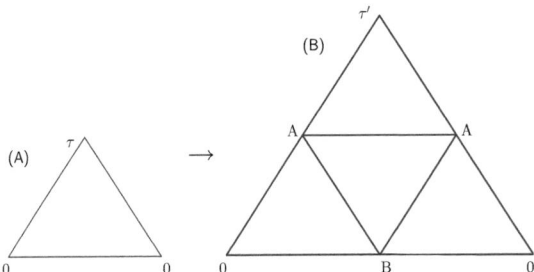

Figure 5.11: **Transit time on the Sierpiński gasket** - Renormalization allows the diffusion dimension of the Sierpiński gasket to be calculated analytically. All vertices are denominated by the time needed for proceeding to the bottom vertices 0 and exiting. *(A)*: The walker enters the triangle at the top vertex and it takes him τ to exit through the bottom vertices. *(B)*: Rescaled triangle, $\tau \to \tau'$.

The transit time of the triangle, τ, is the time the walker needs upon entering at the top vertex to exit through one of the bottom 0 vertices, Fig. 5.11. In Γ_2, the walker needs a time τ to reach the intermediate nodes, from which it takes him the time A to exit. Hence, $\tau' = \tau + A$. From the A-vertices, the walker has a 25% chance to leave Γ_2 after τ, the chance to go to the top vertex is also 25%, as is the chance jump to the other A-vertex or to the B-vertex. All these jumps

5.4 Diffusion on networks

take a typical time τ. Therefore, the time to exit from either of them is

$$A = \frac{1}{4}(0+\tau) + \frac{1}{4}(\tau'+\tau) + \frac{1}{4}(A+\tau) + \frac{1}{4}(B+\tau). \tag{5.69}$$

With the same reasoning, the time to exit from the B vertex is typically $B = (\tau+0)/2 + (\tau+A)/2$. The following renormalization scheme is found

$$\tau' = \tau + A \tag{5.70}$$
$$4A = 4\tau + B + A + \tau' \tag{5.71}$$
$$2B = 2\tau + A. \tag{5.72}$$

The solution of this scheme is $\tau' = 5\tau$ (with $A = 4\tau$ and $B = 3\tau$). The typical transient time is multiplied by 5 each time the side length of the Sierpiński gasket is doubled. The diffusion dimension follows as $d_w = \log 5/\log 2 \approx 2.32$, corresponding to an MSD exponent $\alpha \approx 0.86$. The exact result for the Sierpiński gasket coincides with the simulation results of (finite) Sierpiński triangles, as illustrated in Fig. 5.10.

The typical distance a random walker is displaced in the embedding space after N steps is $R \sim N^{1/d_w}$. The volume in the embedding space in which the walker is to be found, scales as $V \sim R^{d_f} \sim N^{d_f/d_w}$. The probability of the walker being after N steps at the origin is inversely proportional to the volume, i.e.,

$$P(N) \sim N^{-d_f/d_w} = N^{-d_s/2}. \tag{5.73}$$

Here, the spectral dimension of the network is introduced as $d_s = 2d_f/d_w$ [196]. Then, the recurrence probability is $P(N) \sim N^{-d_s/2}$.

5.4.4 Transition networks as models of the configuration space

Recently, network representations of the configuration space are a subject of considerable interest [200]. A natural approach to the dynamics induced by the energy landscape is the location of "basins of attraction" [201], i.e., the regions consisting of the points which lead to the same local minimum if the steepest descent path is followed. Closely connected is the notion of metastable states, *i.e.*, regions of the configuration space which have an extremely long mean escape time. The idea to express the behavior of a dynamical system in terms of the metastable states was first discussed in the context of glass-forming liquids [202, 203]. The role of metastable states in the dynamics of biomolecules was analyzed in [204, 205], specific clustering methods based on the kinetic rather than on the structural similarity were introduced in [206, 207]. Frustration, *i.e.*, the presence of many, nearly isoenergetic local minima, is a typical feature of glassy dynamics [208]. The network of the local minima and transition states has been determined to be a scale free network, *i.e.*, a network whose degree distribution exhibits a power-law tail [209]. Further properties of the networks representing the energy landscape have been discussed, such as clustering, neighbor connectivity, and other topological features [210, 211]. Protein folding is studied in the framework of network theory [195, 212]. Also, network approaches were suggested as a way to find adequate reaction coordinates without the restrictions imposed by projection of the trajectory to low-dimensional subspaces [213]. Minimum-cut techniques are applied to the networks to extract low-dimensional representations of the full dynamics [214–216]. Hierarchical organization of the basins of the energy landscape were reported [206, 209, 213, 217, 218].

ϑ [ps]	1	10	100	1000
$(GS)_2W$	6.1	6.8	7.4	7.9
$(GS)_3W$	6.2	6.8	7.6	8.7
$(GS)_5W$	6.4	6.6	7.0	7.7
$(GS)_7W$	6.1	6.3	6.4	7.3
β-HP, expl. wat.	5.1	5.3	5.7	6.2
β-HP, Langevin	4.6	4.8	5.0	5.6
β-HP, GB/SA	5.4	5.6	5.9	6.5

Figure 5.12: **Fractal dimension** d_f - *The fractal dimension of the network representing the configuration space, $\tilde{\mathbf{S}}(\vartheta)$. The network is constructed from the MD trajectory projected to a subset of PCs, most often the first ten PCs. Eq. (5.65) allows the fractal dimension to be calculated from the volume scaling of the network. None of the network fills the full ten-dimensional embedding space, all networks exhibit fractal behavior.*

In the following, the anomalous diffusion seen in the kinetics of the β-hairpin

5.4 Diffusion on networks

and the $(GS)_nW$ peptide simulations [see Sec. 4.3] are analyzed using network representations of the configuration space. First, the volume sampled by the MD trajectory in the configuration space is partitioned into 10 000 discrete states. These discrete states are randomly taken from the frames in trajectory. Here, we divide the simulation length, T, into 10 000 pieces of equal length and take the first frame of each piece as the center of a discrete state, r_i with $i = 1, 2, ..., 10\,000$ being the configuration space coordinates. A frame in the trajectory is now assigned to the discrete state which has the least Euclidean distance in the mass-weighted coordinates. In this way, the entire trajectory $(x_1, x_2, ..., x_T)$ is mapped to a discrete time series, $(\kappa_1, \kappa_2, ..., \kappa_T)$. From the discrete time series the transition matrix, $\tilde{\mathbf{S}}(\vartheta)$ can be obtained using Eq. (5.63). The network represented by the transition matrix is embedded in the configuration space, the r_i being the coordinates of node i. The network representation enables us to study how the energy landscape brings about the subdiffusive dynamics of the molecule. As the network analysis is numerically cumbersome in the full configuration space, we perform the analysis on the subspace spanned by the first ten PCs. These ten strongly delocalized modes account for more than 50% of the overall fluctuations in the molecule. However, with the considerations in Sec. 5.1, the appearance of memory effects above 1 ps is expected.

The fractal dimension of the transition network, $\tilde{\mathbf{S}}(\vartheta)$, is obtained using Eq. (5.65). The number, N of edges enclosed in the sphere of radius R centered at r_i is computed for various R. The function $N(R)$ is averaged over 1 000 nodes, and Eq. (5.65) yields the fractal dimension d_f by a least-squares power-law fit to the data, Tab. 5.12.

Networks corresponding to various ϑ are used as Markov models. A random walk is performed on the network with transition matrix $\tilde{\mathbf{S}}(\vartheta)$. The discrete time evolution generated from the transition matrix is now inversely transformed to the configuration space by replacing the discrete state number, i, with the configuration space coordinates of the state, r_i. Doing so, a the random walk on the network is mapped to a random walk in the configuration space. The random walk dimension, d_w, is obtained from the MSD in the configuration space. It is given together with the MSD exponent α in Tab. 5.14.

The network diffusion reproduces the effect of subdiffusivity; the MSD ex-

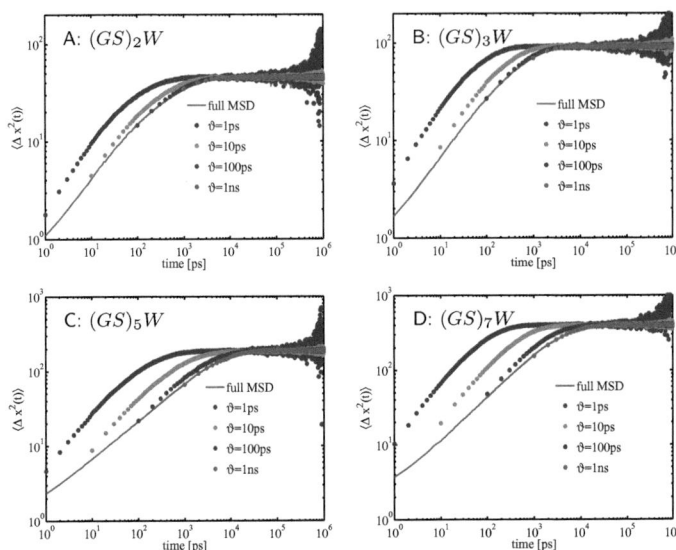

Figure 5.13: *(Color online)* **Diffusion on transition networks.** - *For the $(GS)_nW$ peptides with $n = 2,3,5,7$, a random walk is performed on the networks $\tilde{\mathbf{S}}(\vartheta)$ with various time lag ϑ (see legends). The networks are obtained from the projection of the MD trajectories to the subspace of the first ten PCs. The MSD of the random walk on the different networks is compared to the MSD (full red line) found in the first ten PCs of the original MD trajectory. All networks exhibit subdiffusive dynamics to some extent. However, for the time lags $\vartheta = 1$ ps and 10 ps, the network random walk overestimates the MSD. For $\vartheta = 100$ ps the networks reproduce the subdiffusion of the original MSD sufficiently. The networks for $\vartheta = 1$ ns, the MSD is close to the saturation level. Hence, the saturation conceals possible subdiffusive effects.*

ponent α is in the subdiffusive regime, Tab. 5.14. However, the networks for $\vartheta = 1$ ps and 10 ps usually exhibit an MSD clearly larger than the original trajectory, given as red line in Fig. 5.13. The reason for the difference may be memory effects present on the short time scales, which can arise from the projection to the ten-dimensional subspace of the first ten PCs. The memory effects cannot be reproduced by the Markovian network dynamics. Therefore, the subdiffusivity is underestimated, *e.g.*, the network MSD is larger than the MSD found in the

5.4 Diffusion on networks

	d_w				α			
ϑ [ps]	1	10	100	1000	1	10	100	1000
$(GS)_2W$	2.6	3.0	4.3	11.3	0.8	0.7	0.5	0.2
$(GS)_3W$	2.4	2.7	3.8	9.6	0.8	0.7	0.5	0.2
$(GS)_5W$	2.4	2.7	3.2	4.4	0.8	0.7	0.6	0.5
$(GS)_7W$	2.3	2.5	3.0	4.4	0.9	0.8	0.7	0.5
β-HP, expl. wat.	3.1	3.6	4.5	5.7	0.7	0.6	0.4	0.3
β-HP, Langevin	2.9	3.6	4.5	6.4	0.7	0.6	0.4	0.3
β-HP, GB/SA	3.5	3.8	4.4	6.4	0.6	0.5	0.5	0.3

Figure 5.14: **Diffusion dimension d_w and MSD exponent α** - A random walk on the networks representing the configuration space, $\tilde{\mathbf{S}}(\vartheta)$, is performed. The network is obtained from the projection of the trajectory to the first ten PCs. The random walk on the network is inversely transformed to the configuration space. The diffusion dimension d_w is calculated from the MSD using Eq. (5.64). The MSD exponent, α, used throughout the present thesis to quantify the subdiffusivity, is given in the right panel for comparison with the results of Chap. 4. All networks exhibit subdiffusive dynamics. The network $\tilde{\mathbf{S}}(\vartheta)$ with $\vartheta = 1$ ns is in all cases close to the saturation level. Therefore, the exponents tend to be underestimated as a consequence of finite size effects. For $\vartheta = 100$ ps the MSD exponent is close to the values found in the MD simulations, see Fig. 4.14. The networks for $\vartheta = 1$ ps and 10 ps tend to overestimate the MSD exponent. This is due to the presence of memory effects arising from the projection to the first ten PCs and the restricted spatial resolution, which becomes important for the dynamics on the shortest time scales. Furthermore, the networks of the β-hairpin simulations may be affected by the poor sampling due to the non-convergence, see Sec. 4.3.

MD simulation data. Also, the limited spatial resolution of a network with 10 000 vertices affects the MSD, and introduces in additional noise, in particular on the short time scales. For $\vartheta = 100$ ps, the network MSD and the original MSD exhibit an acceptable coincidence. The significance of the networks with $\vartheta = 1$ ns is limited by the fact that the time resolution is close to the time scale of saturation, on which the walker is affected by the finite, accessible volume. The finite volume also accounts for the tendency of the MSD exponent α to be underestimated by the random walks on the 1 ns transition network, $\tilde{\mathbf{S}}(\vartheta = 1\text{ns})$.

For completeness, the spectral dimension is obtained using Eq. (5.73), see

Tab. 5.15. It is most likely a finite size effect that, in contrast to the argument employed in the discussion of Eq. (5.73), $d_s \neq 2d_f/d_w$.

ϑ [ps]	1	10	100	1000
$(GS)_2W$	0.7	0.8	1.2	4.3
$(GS)_3W$	0.6	0.6	0.8	2.5
$(GS)_5W$	1.0	0.9	0.8	0.9
$(GS)_7W$	1.1	0.9	0.9	0.7
β-HP, expl. wat.	1.2	1.2	1.3	1.3
β-HP, Langevin	1.1	1.1	1.2	1.4
β-HP, GB/SA	1.2	1.1	1.1	1.3

Figure 5.15: **Spectral dimension** d_s - The spectral dimension of the networks representing the configuration space, $\tilde{\mathbf{S}}(\vartheta)$. The network is obtained from the MD trajectories projected to the first ten PCs. Eq. (5.73) yields d_s using the recurrence probability. The values of d_s do not obey to $d_s = 2d_f/d_w$, potentially due to the finite size of the networks with only 10 000 nodes.

Influence of projection

In the above analysis, the dynamics were projected to the subspace spanned by the first ten PCs, *i.e.*, the network is embedded to a d-dimensional embedding space with $d = 10$. In order to characterize the influence of the projection, the dynamics of the $(GS)_5W$ peptide is also analyzed for projections to subspaces with $d = 1, 3$, and 100. The fractal dimension of the networks is given in Tab. 5.17.

The networks obtained from the projections to $d = 1$ and $d = 3$ dimensions are in the range 1.4–1.7. The values obtained for $d = 10$ are close to those d_f values found for $d = 100$. Hence, the fractal structure of the network $\tilde{\mathbf{S}}(\vartheta)$ is essentially developed in the subspace spanned by the first ten PCs. However, the MSD of the network, $\tilde{\mathbf{S}}(\vartheta)$, obtained from the projection to $d = 100$ dimensions is considerably closer to the original MSD than the one for $d = 10$. The fact that the $d = 10$ network exhibits the same fractal exponent as the one for $d = 100$ but is inferior in the reproduction of the kinetics demonstrates that the higher PC modes (here the PC modes 11 to 100) contain only a small contribution to the

5.4 Diffusion on networks

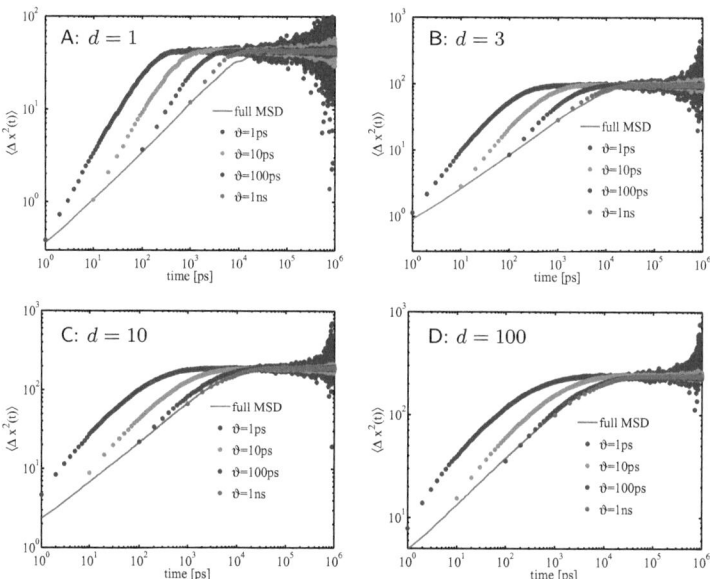

Figure 5.16: (Color online) **Projection of $(GS)_5W$ to configurational subspaces – network diffusion in various dimensions.** - The full MD trajectory was projected to d-dimensional subspaces of the configurational space, with $d = 1$, 3, 10, and 100. From the projected trajectory a transition matrix, $\tilde{\mathbf{S}}(\vartheta)$, was obtained with lag time $\vartheta = 1\,\mathrm{ps}$, $10\,\mathrm{ps}$, $100\,\mathrm{ps}$, and $1\,\mathrm{ns}$. A random walk on the network defined by $\tilde{\mathbf{S}}(\vartheta)$, is inversely transformed to the configuration space and the random walker's MSD calculated. The transition matrices obtained from the projections to low dimensional spaces exhibit a larger difference to the MSD of the original MD trajectory, given as red, full line. For $d = 100$ the transition networks, $\tilde{\mathbf{S}}(\vartheta)$, with $\vartheta = 100\,\mathrm{ps}$ and $1\,\mathrm{ns}$ reproduce the original kinetics with high accuracy, but also for $\vartheta = 1\,\mathrm{ps}$ and $10\,\mathrm{ps}$ the difference to the original MSD is considerably smaller than for the networks obtained from projections to subspaces of lower dimension.

overall fluctuations, but are kinetically not insignificant. The differences between the network MSD and the MSD obtained from the original MD trajectory has three origins: (i) memory effects arising from the neglected dimensions, *i.e.*, due to the non-Markovian nature of the trajectory (ii) limited spatial resolution, and (iii) memory effects rooted in the discretization scheme.

ϑ [ps]	1	10	100	1000		1	10	100	1000
d	d_f				d	α			
1	1.6	1.6	1.7	1.7	1 (0.5)	0.9	1.0	0.8	0.6
3	1.4	1.4	1.5	1.6	3 (0.5)	0.9	0.9	0.8	0.5
10	6.4	6.6	7.0	7.7	10 (0.5)	0.8	0.7	0.6	0.5
100	6.3	6.5	6.8	7.7	100 (0.4)	0.7	0.6	0.5	0.3

Figure 5.17: **Projection of** $(GS)_5W$ **to configurational subspaces – fractal dimension of transition networks.** - The full MD trajectory was projected to d-dimensional subspaces of the configurational space, with $d = 1, 3, 10$, and 100. From the projected trajectory a transition matrix, $\tilde{\mathbf{S}}(\vartheta)$, was obtained with lag time $\vartheta = 1\,\mathrm{ps}, 10\,\mathrm{ps}, 100\,\mathrm{ps}$, and 1 ns. The fractal dimesion of the networks is obtained using Eq. (5.64). The fractal dimension found for $d = 10$ is close to the values obtained from $d = 100$, i.e., the fractal shape of the network is mainly contained in the first ten PCs. The MSD exponent, α, is calculated from the random walks illustrated in Fig. 5.16.

The above results demonstrate that the configuration space of a molecule can be represented by a network model. A Markov process on the network can reproduce the kinetics of the molecule. Subdiffusive dynamics as seen by the sublinear time-dependence of the MSD arise from a fractal-like geometry of the network. The representation of the kinetics is limited by three findings. Memory effects due to the projection of the dynamics are present and violate the Markov property on short time scales. In order to be numerically tractable and to have sufficient statistics, the number of discrete states is limited and restricts the spatial resolution of the network in the configuration space. Finally, the discretization is not uniquely defined. If kinetically different structures in the configuration space are mapped on the same network vertex, the transition paths are unnaturally mixed. The mixing disturbs the fractal geometry and makes the network more permeable, *i.e.*, the MSD is larger than the MSD of the original trajectory.

5.4.5 Eigenvector-decomposition and diffusion in networks

Since random walks on networks can be modeled as Markov processes based on a transition matrix, methods from linear algebra can serve as useful tools in the

5.4 Diffusion on networks

analysis of diffusion problems. The time evolution of a Markov process can be expressed in terms of the eigenvectors of the transition matrix. The diffusive contribution of each eigenvector can be quantified with its transport coefficient. Only a small fraction of eigenvectors dominate the MSD seen for the random walker of a given network.

Let $\mathbf{S}$ be an irreducible[6] transition matrix and $\boldsymbol{p}(n)$ the probability vector after n time steps. Its components $p_i(n)$ correspond to the probability of being in state i after n time steps. The initial probability vector is $\boldsymbol{p}(0)$. The time evolution of the probability $\boldsymbol{p}$ on the network is governed by the following equation

$$\boldsymbol{p}(n+1) = \mathbf{S}\boldsymbol{p}(n). \tag{5.74}$$

The recursive Eq. (5.74) yields the evolution equation

$$\boldsymbol{p}(n) = \mathbf{S}^n \boldsymbol{p}(0). \tag{5.75}$$

The largest eigenvalue of the matrix $\mathbf{S}$ equals one, $\lambda_1 = 1$, and has multiplicity one, *i.e.*, there is a uniquely defined eigenvector to the eigenvalue $\lambda_1 = 1$, according to the Ruelle-Perron-Frobenius theorem [192]. The theorem also states that all eigenvalues have an absolute value smaller than one, $|\lambda_k| < 1$ for all $k \neq 1$. Furthermore, the eigenvector to $\lambda_1 = 1$ is the only eigenvector with strictly non-negative components. The eigenvector to the largest eigenvalue, $\lambda_1 = 1$, is the stationary probability vector $\boldsymbol{p}^s$ and it is

$$\lim_{n\to\infty} \boldsymbol{p}(n) = \lim_{n\to\infty} \mathbf{S}^n \boldsymbol{p}(0) = \boldsymbol{p}^s. \tag{5.76}$$

The sum of the components of the stationary eigenvector $\boldsymbol{p}^s$ equals one, $\sum_k p_k^s = 1$. From Eq. (5.74) it follows, that the component sum of any other eigenvector $k \neq 1$ equals zero, *i.e.* $\sum_k p_k = 0$ if $k \neq 1$.

---

[6] The graph of a network is irreducible if any vertex can be reached from every other vertex, *i.e.* the network cannot be decomposed into different, unconnected subgraphs. The unconnected subgraphs of a reducible graph can be treated separately, each subgraph being irreducible. The corresponding transition matrix is called irreducible if it is not similar to a block upper triangular matrix via a permutation. The definitions of irreducibility of the network and the transition matrix are equivalent [192].

The probability $p(n)$ can be seen as a diffusion process, based on a large number (an ensemble) of random walkers. Eq. (5.74) can be modified to

$$p(n+1) - p(n) = (\mathbf{S} - \mathbb{1})p(n). \tag{5.77}$$

Eq. (5.77) is the discrete analog of the diffusion equation, Eq. (2.33). The operator $(\mathbf{S}-\mathbb{1})$ is the discrete Laplace operator and has the eigenvalues $\{\lambda_i - 1 | i = 1, 2, ...\}$. In the continuous time limit, the $\mu = -\log \lambda \approx -(\lambda - 1)$ are the (negative) eigenvalues of the Laplace operator. The eigenvectors of the Laplace operator are identical with the eigenvectors of $\mathbf{S}$. From Eq. (5.77) it follows that the eigenfunctions of the Laplace operator decay exponentially with the characteristic time $1/\mu$, except the eigenfunction with the eigenvalue $\mu = 0$, i.e., corresponding to the equilibrium p^s with $\lambda = 1$.

We introduce the notation $(\operatorname{diag} \boldsymbol{v})_{ij} = \delta_{ij} v_j$ for the components of the matrix $\operatorname{diag} \boldsymbol{v}$, where $\boldsymbol{v}$ is a vector. As pointed out earlier, in general, the transition matrix, $\mathbf{S}$, is not symmetric. From the detailed balance condition it follows

$$s_{ij} p_j^s = s_{ij} p_i^s \quad \text{for all } i, j. \tag{5.78}$$

Therefore, the matrix

$$\tilde{\mathbf{S}} = \operatorname{diag}(\boldsymbol{p}^s)^{-1/2} \mathbf{S} \operatorname{diag}(\boldsymbol{p}^s)^{1/2} \tag{5.79}$$

is symmetric, i.e. $\tilde{\mathbf{S}} = \tilde{\mathbf{S}}^T$, where $\tilde{\mathbf{S}}^T$ is the transposed of the matrix $\tilde{\mathbf{S}}$.

As $\tilde{\mathbf{S}}$ is symmetric, it has real eigenvalues $\{\lambda_k\}$ and the eigenvectors, $\{\tilde{\boldsymbol{v}}_k\}$, obey

$$\tilde{\mathbf{S}} \tilde{\boldsymbol{v}}_k = \lambda_k \tilde{\boldsymbol{v}}_k, \tag{5.80}$$

and form an orthogonal basis set. The diagonal matrix that is similar to $\tilde{\mathbf{S}}$ can be expressed as

$$\mathbf{U}^t \tilde{\mathbf{S}} \mathbf{U} = \mathbf{\Lambda}. \tag{5.81}$$

The eigenvectors of $\mathbf{S}$ are obtained from the $\{\tilde{\boldsymbol{v}}_k\}$ as $\boldsymbol{v}_k = \operatorname{diag} \boldsymbol{p}^s \tilde{\boldsymbol{v}}_k$. The transition matrix $\mathbf{S}$ has the same eigenvalues, λ_k, as the matrix $\tilde{\mathbf{S}}$. However, the eigenvectors of $\mathbf{S}$ do, in general, not form an orthogonal set. Furthermore, the identity

$$\mathbf{S} = \operatorname{diag}(\boldsymbol{p}^s)^{1/2} \tilde{\mathbf{S}} \operatorname{diag}(\boldsymbol{p}^s)^{-1/2} \tag{5.82}$$

5.4 Diffusion on networks

allows together with
$$\operatorname{diag}(\boldsymbol{p}^s)^{-1/2}\operatorname{diag}(\boldsymbol{p}^s)^{1/2} = \mathbb{1} \tag{5.83}$$
the powers of the transition matrix to be expressed as
$$\mathbf{S}^n = \operatorname{diag}(\boldsymbol{p}^s)^{1/2}\tilde{\mathbf{S}}^n\operatorname{diag}(\boldsymbol{p}^s)^{-1/2}. \tag{5.84}$$
Using Eq. (5.84), the time evolution Eq. (5.75) can be written as
$$\boldsymbol{p}(n) = \operatorname{diag}(p^s)^{1/2}\mathbf{U}\boldsymbol{\Lambda}^n\mathbf{U}^t\operatorname{diag}(p^s)^{-1/2}\boldsymbol{p}(0). \tag{5.85}$$
When using the componentwise notation, Eq. (5.85) reads
$$p_i(n) = \sum_{kl}(p_i^s)^{1/2}u_{ik}\lambda_k^n u_{lk}(p_l^s)^{-1/2}p_l(0). \tag{5.86}$$
The conditional probability of being in state i after n steps starting in state q at time $t=0$ is
$$p_i(n|q,0) = \sum_{kl}(p_i^s)^{1/2}u_{ik}\lambda_k^n u_{lk}(p_l^s)^{-1/2}\delta_{lq} \tag{5.87}$$
$$= \sum_k (p_i^s)^{1/2}u_{ik}\lambda_k^n u_{qk}(p_q^s)^{-1/2}. \tag{5.88}$$
The joint probability of being in state i at step n and in state q at time 0 reads
$$p_i(n;q,0) = p_q(0)p_i(n|q,0) = p_q(0)\sum_k (p_i^s)^{1/2}u_{ik}\lambda_k^n u_{qk}(p_q^s)^{-1/2}. \tag{5.89}$$
Let d_{iq} be the distance between state i and state q in the embedding space, $d_{iq} = \|\boldsymbol{r}_i - \boldsymbol{r}_q\|$. Then, the MSD of the random walk on the network defined by $\mathbf{S}$ is given as
$$\langle \Delta x^2(n) \rangle = \sum_{iq} d_{iq}^2 p_i(n;q,0) = \sum_{iq} d_{iq}^2 p_q(0)\sum_k (p_i^s)^{1/2}u_{ik}\lambda_k^n u_{qk}(p_q^s)^{-1/2}. \tag{5.90}$$
Note that the above expression is an ensemble average. It can be performed with any given initial probability distribution $\boldsymbol{p}(0)$. In equilibrium, one has to replace $\boldsymbol{p}(0)$ with the stationary distribution $\boldsymbol{p}^s$, so one has
$$\langle \Delta x^2(n) \rangle = \sum_{iq} d_{iq}^2 p_i(n;q,0) = \sum_{iq} d_{iq}^2 p_q^s \sum_k (p_i^s)^{1/2}u_{ik}\lambda_k^n u_{qk}(p_q^s)^{-1/2}. \tag{5.91}$$

Defining the *transport coefficients* as

$$R_k := \sum_{iq} d_{iq}^2 (p_q^s)^{1/2} (p_i^s)^{1/2} u_{ik} u_{qk}, \tag{5.92}$$

Eq. (5.91) can be expressed as

$$\langle \Delta x^2(n) \rangle = \sum_k R_k \lambda_k^n. \tag{5.93}$$

The long-time limit of the MSD, $n \to \infty$, is given by the largest eigenvalue $\lambda_1 = 1$ and its transport coefficient R_1, that is, $\lim_{n \to \infty} \langle \Delta x^2(n) \rangle = R_1$. All other eigenvalues have an absolute value smaller than one. Therefore, the contribution of these eigenvalues to the MSD decays for $n \to \infty$, except the coefficient R_1. As the MSD is strictly monotonic increasing, it follows $R_k < 0$ for $k \neq 1$. The larger eigenvalues, *i.e.* the ones close to one, contribute the long-time behavior of Eq. (5.93). If the sum in Eq. (5.93) is performed only over the low-indexed mode numbers, *i.e.* over the largest eigenvalues, then the short-time behavior will not be reproduced correctly. As $\langle \Delta x^2(n=0) \rangle = 0$, it follows $\sum_k R_k = 0$. Therefore $R_1 = -\sum_{k>1} R_k$.

The MSD can be expressed as

$$\langle \Delta x^2(t) \rangle = \sum_k R_k e^{t \log \lambda_k} = \sum_k R_k e^{-\mu_k t}, \tag{5.94}$$

with the rates defined as $\mu_k = -\log \lambda_k > 0$ and the continuous time coordinate t instead of the discrete n. The MSD can be decomposed into a sum of exponential contributions with a typical time μ_k^{-1}. The prefactors are the transport coefficients, R_k, with the properties given above.

The MSD of a Markov process in Eq. (5.94) gives rise to subdiffusion if the transport coefficients, R_k, obey to certain conditions, as briefly outlined in Appendix E. It turns out that only a small number of additive terms suffice to provide a power-law MSD over several time scales. Therefore, even the Sierpiński triangles, Γ_n, for $n \geqslant 3$ exhibit a noticeable subdiffusivity, as illustrated in Fig. 5.10.

As an example, we analyze the eigenvalues of the transition matrix corresponding to the Sierpiński triangles. The eigenvalues of the discrete Laplace operator are illustrated in Fig. 5.19 A. The eigenvalues found for Γ_n are also found for Γ_k

5.4 Diffusion on networks

with $k > n$, but with a higher multiplicity. The eigenvalues exhibit two series of multiplicities, one with $2, 3, 6, ... (3^k+3)/2$ and the other with $1, 4, 13, (3^k+3)/2-2$ [219, 220].

The transport coefficients, R_k, in Eq. (5.93) determine the MSD seen in the Sierpiński triangle Γ_n. The first exponent, R_1, contributes the saturation plateau asymptotically reached in the limit $t \to \infty$. All other transport coefficients are negative. The transport coefficients R_k (for $k \neq 1$) are largest for the two eigenvalues $k = 2$ and 3. It turns out that only isolated transport coefficients give a significant contribution to the MSD. The decomposition of the MSD into the different exponential contributions, Eq. (5.94), is illustrated in Fig. 5.18. Note the similarity with Fig. E.1.

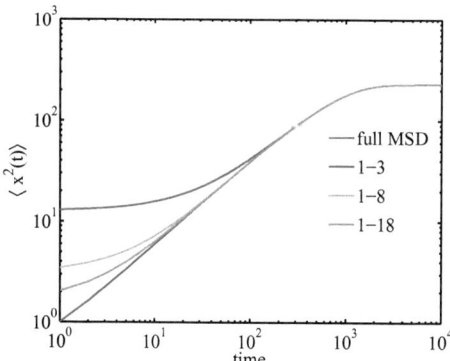

Figure 5.18: (Color online) **MSD of the Sierpinski gasket** Γ_6 - The red line illustrates the complete MSD, while the other curves correspond to Eq. (5.93) with only a partial sum over the coefficients R_k, over the first three (magenta), over the first eight (cyan) and over the first 18 (green) eigenvalues, see legend. The $R_k > 0$ of the largest eigenvalues contributes the value of the plateau reached for long t. With decreasing λ_k, i.e. with decreasing μ_k and increasing k, the short-time behavior is retained. Only a small number of transport coefficients determine the behavior of the MSD up to high accuracy.

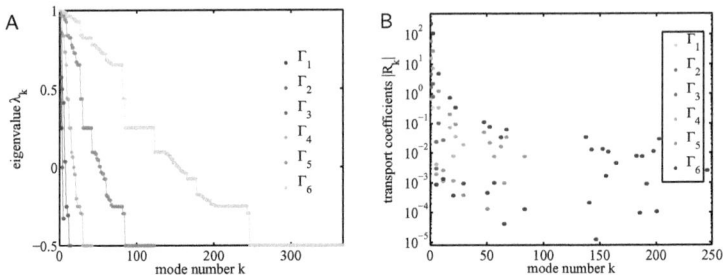

Figure 5.19: (Color online) **Diffusion on the Sierpiński triangle** - A: Eigenvalues of Sierpiński triangles with $n = 1, 2, 3, 4, 5,$ and 6. B: Absolute value of transport coefficients, $|R_k|$, for various Sierpiński triangles. Only a limited number of the R_k has an absolute value significantly larger than zero. The largest $|R_k|$, apart from R_1, is seen for the eigenvalues 2 and 3, 7 and 8, 17 and 18.

5.5 Conclusion

Different approaches have been applied to model subdiffusion in the internal dynamics of biomolecules. It has been demonstrated with Zwanzig's projection formalism that projections to a limited set of relevant coordinates gives rise to correlations and memory effects. From the simulations presented in Chap. 4 it follows that the memory effects due to the projection to the internal coordinates of a β-hairpin molecule on the time scale of 2 ps are not significant for the kinetic behavior, in particular for the MSD found.

The Rouse chain model is briefly reviewed. It is demonstrated how the normal modes of Rouse model determine the time evolution. The distance fluctuations of the Rouse chain are shown to exhibit subdiffusion. However, an estimation of the Rouse time demonstrated that the subdiffusive regime does not fit to the configurational subdiffusion seen in MD simulations and in single molecule spectroscopy.

The CTRW model has been introduced and was found to not reproduce subdiffusivity in the time-averaged MSD. We carefully analyzed the confined CTRW model, *i.e.*, the CTRW in a finite volume. Although the confined CTRW does

5.5 Conclusion

exhibit subdiffusive dynamics in the time-averaged MSD, it is not an explanation of the subdiffusion seen in the configurational diffusion of the MD simulations at hand.

The configuration space can be represented by a discrete transition network. It was demonstrated that a Markov process based on the network representation of the configuration space allows the subdiffusive MSD of the MD simulation to be reproduced with high accuracy on time scales of and above 100 ps. The fractal-like nature of the transition network was characterized by the fractal dimension and identified as the essential mechanism that gives rise to subdiffusive dynamics.

It is sketched that, in principal, the transition matrices allows methods from linear algebra and matrix algebra to be applied to the problem of subdiffusion. The transport coefficients arising from the eigenvectors of the transition matrix determine the diffusional behavior of the network. However, as being numerical cumbersome, these methods could not yet be applied to the transition networks obtained from the MD trajectories.

> *[...] mais n'abandonnez jamais. A la longue votre article sera accepté et publié dans le Physical Review et alors, [...], vous serez devenu un vrai physicien.*
>
> DAVID RUELLE

CHAPTER 6
CONCLUDING REMARKS AND OUTLOOK

6.1 Conclusion

The present thesis deals with questions from two very active fields of physical research, the field of protein dynamics and the topic of subdiffusive transport processes.

The cell is the building block of terrestrial life. An understanding of cell activity, in turn, requires the understanding of proteins, the main agents at the sub-cellular level. Therefore, the study of protein dynamics has attracted an enormous, still increasing attention in the last decades. The experimental techniques provide access to details of the molecular processes with unprecedented accuracy. The entanglement of processes on very different time and length scales puzzles the theoreticians engaged in the field of biophysics.

Even on the single-molecule level, the processes span many orders of magnitude. In the case of peptides, the fastest motion are due to covalent bond vibrations, found in the range of femtoseconds (10^{-15} s) to large domain motions, which occur on the time scale of some hundreds of microseconds (10^{-1} s). Hence, internal dynamics of peptides span at least 14 orders of magnitude.

In order to give a coherent description of the very different time scales on which the peptide and protein dynamics occur, the energy landscape picture was successfully employed since the 1970s [22, 195, 221, 222]. Notably the energy landscape idea has been introduced in the context of equilibrium dynamics [23], albeit energy landscape models also proved extremely fruitful for the analysis of protein folding.

The energy landscape idea has been extensively studied in the context of glass forming liquids. In contrast to liquids, proteins fold into a relatively well defined conformation, the *native state*. However, the similarity of proteins and peptides to glassy systems is due to the presence of different isoenergetic substates (for biomolecules, found in the same native state), which are separated from each other by energetic barriers. That is, like glass formers, peptides exhibit *frustration*. The substates themselves may be further structured by even lower internal barriers. Hence, the molecule's energy landscape is organized by a *hierarchy of states* allowing a vast range of time and length scales to be covered.

In the thesis at hand, the thermal fluctuations of biomolecules have been studied using MD simulations. The kinetics of four different, short peptide chains and a β-hairpin molecule has been examined on the basis of MD trajectories extending to the microsecond time scale. In the introduction chapter, three questions are raised. Here, the results obtained in the present thesis shall be summarized in response to these questions.

Is subdiffusivity a general feature of internal fluctuations in biomolecules?

Subdiffusive internal fluctuations have been found in experiments with myoglobin [23, 30, 50], lysozyme [54] and flavin oxireductase [27, 28, 51, 52] and simulations of lysozyme [53], the outer membrane protein f [136], and flavin [183]. Here, the MD simulations of short peptide chains allows the subdiffusivity of internal biomolecular motion to be assessed for small and relatively simple systems. The fact that the $(GS)_n W$ peptides for all $n = 2, 3, 5,$ and 7 exhibit subdiffusive fluctuations corroborates the generality of the effect. Also, the β-hairpin undergoes fractional diffusion. A well defined, native state is not a prerequisite of anomalous diffusion in the internal coordinates. Therefore, fractional dynamics are likely to be the standard case rather than an exception or the property of specific, more complex molecules.

6.1 Conclusion

The comparison between different models for the water dynamics has indicated shortcomings of the generalized Born approach (GB/SA) with respect to the kinetic behavior, *i.e.*, while the free energy profile is correctly reproduced the kinetics are not. Although the effect of subdiffusion is present in the GB/SA simulation, the MSD exponent is considerably different.

What is the mechanism that causes the internal dynamics of biomolecules to be subdiffusive?

Can the CTRW model exhibit subdiffusivity in the time-averaged MSD?

Hitherto, several models have been suggested to describe the experimental and simulation findings [27, 28, 53, 54, 155, 183]. In this thesis, we were seeking the underlying mechanism of subdiffusivity, so we examined only approaches that make explicit statements about the origins of the effect. We did not address *ad hoc* descriptions employed to derive further molecular properties from fractional diffusion.

The harmonic chain model, *i.e.* the Rouse chain, and its extension to more general geometries [151, 155, 156], although exhibiting subdiffusive behavior, has been found to fail in the reproduction of the correct time-scales of the fractional regime. The CTRW model has been found to have a subdiffusive ensemble average. However, the simulation results and some of the experiments exhibit subdiffusion in the time-averaged MSD. Therefore, the time averages of CTRW quantities have been examined carefully. The free, unbounded CTRW does not display a subdiffusive, time-averaged MSD. In contrast, the time-averaged MSD of the confined CTRW exhibits a subdiffusive behavior. These findings make the CTRW a candidate for the mechanism responsible for the anomalous diffusion found in biomolecules. However, an estimation of the parameters of the confined CTRW, in particular the critical time which defines a lower bound for the presence of subdiffusion in the time domain, disqualifies CTRW as a model for biomolecular kinetics.

As an alternative approach, the diffusion on fractal networks has been discussed. A transition network representation of the configuration space arises naturally by dividing the energy landscape in discrete states. The network geometry was analyzed and found to be of fractal nature. The Hausdorff dimension of the transition network was calculated and lies in the range $d_f \approx 6$ to 7. The transition networks obtained from the MD simulation trajectories were also used as an input for Markov models, which allowed random walks on the networks to be performed. The transition network random walks were used to determine the diffusion dimension. The dimensions found clearly indicate subdiffusive behavior. Also the MSDs, as predicted by the transition networks, were obtained and compared to the results of the original MD trajectories. The kinetics are correctly reproduced at and above the time scale of 100 ps. For shorter times, the kinetics are not completely represented, although subdiffusive behavior has still been found. Three potential sources of deviations have been identified: (i) the dynamics of the projected trajectory is effectively non-Markovian on short time scales; (ii) the spatial resolution of the network is too low; and (iii) the discretization scheme is too simple to resolve the correct kinetics. However, the memory effects accounting for the non-Markovian nature have been found to be insubstantial, as was demonstrated by a comparison of an explicit water simulation to a Langevin thermostat simulation of the β-hairpin. Both these simulations agree in terms of the overall kinetic behavior. It was illustrated that the first ten PCs account for most of the internal fluctuations. However, the low amplitude, high-indexed PCs yield important dynamical and kinetic information. Network models taking the higher PCs into account are much more accurate in reproducing the biomolecular dynamics. Briefly, we sketched how algebraic methods may be applied in the context of transition networks to give a better understanding of subdiffusive behavior.

The fractal geometry of the accessible volume in the configuration space is a consequence of the energy landscape. A high potential energy makes configurations unfavorable such that they are effectively forbidden. Hence, regions with too high a potential energy value are effectively unaccessible. The configurations observed in the MD trajectory sample the accessible volume, which is divided by multiple barriers into configurational substates. The dynamics are further restricted by geometrical constraints. That is, the accessible volume is not compact but contains many unaccessible "islands" which act as obstacles and render the accessible volume fractal-like.

Fractal behavior prompts to scale invariance. The power-law behavior of the

MSD, $\langle \Delta x^2(t) \rangle \sim t^\alpha$, means that the dynamics on the time scale of, say $t = 1\,\mathrm{ps}$, are effectively the same as those on the time scale of $t' = 1\,\mathrm{ns}$, provided the distances are scaled as $l' = \lambda^{\alpha/2} l$ with $\lambda = t'/t$. Therefore, the subdiffusive MSD characterizes an ordering principle of the energy landscape in a specific range.

The kinetics of the peptides studied in the present thesis cover the range from picoseconds to microseconds. As the molecules studied are relatively small with a simple structure, this time range is sufficient to collect enough statistics for a reliable analysis. At least in the case of the $(GS)_n W$ peptides, we expect the longest intrinsic time scales to be in the range of $10\,\mathrm{ns}$. For larger systems, much longer time scales will be found. However, we expect similar dynamical behavior, *i.e.* subdiffusivity, to be present on intermediate time scales even for much larger biomolecules.

6.2 Outlook

The present work provides a new approach in the analysis of internal fluctuations of biomolecules. Well established network methods may be a key to quantifying the properties of the energy landscape which give rise to subdiffusive behavior. However, it is an open question how to discretize the energy landscape in order to reproduce the dynamics on shorter time scales correctly, in particular with respect to subdiffusivity. Network models likely allow the organizing principles of the energy landscape to be characterized. The hierarchical organization of the energy landscape, as reported in recent publications [206, 209, 213, 217, 218], could be the basis to establishing a rigorous length-time scaling that determines the MSD exponent α. If function-relevant subgraphs can be identified, the analysis of network representations of the configuration space can give new insights in the connection between dynamics and biological function [204, 206, 223].

A minimal mechanical model exhibiting subdiffusive fluctuations is needed to assess which parts of the interaction potentials used in MD simulations account for the fractional behavior. In the present work, we did not study the temperature dependence of the subdiffusive behavior. Recent experiments demonstrated small peptides to exhibit a dynamical transition [48], as earlier found for proteins [31].

An interesting question is whether the subdiffusive behavior is lost below the glass transition when the molecule is expected to be trapped in a single potential well. Also the precise time dependence of the MSD at lower temperatures could tell more about the hierarchical nature of the energy landscape.

A further challenge is to derive quantities from the theoretical models which can be assessed by experimental techniques. The presence of fractional diffusion on the pico- and nanosecond time scale has been shown to give rise to ACFs of Mittag-Leffler form [54], regardless of the underlying mechanism. Therefore, it is difficult to put the theoretical models to the test as long as only the ACFs can be measured, such as in spin echo neutron scattering or single molecule fluorescence spectroscopy. Translating a given transition network model into neutron scattering quantities is a promising path to quantify the contributions of the elementary dynamical processes to the dynamical structure factor, the intermediate scattering function, and other measurable quantities.

REFERENCES

[1] R. BROWN. A brief account of microscopical observations on particles contained in the pollen of plants. *Phil. Mag.*, **4** (1828):161–173. 1, 11

[2] A. FICK. Ueber Diffusion. *Pogg. Ann. d. Phys. u. Chem.*, **44** (1855) 1:59–86. 1, 9

[3] A. EINSTEIN. Über die von der molekularkinetischen Theorie der Wärme geforderte Bewegung von in ruhenden Flüssigkeiten suspendierten Teilchen. *Ann. Phys. (Leipzig)*, **17** (1905):549. 1, 11, 24

[4] J. PERRIN. Mouvement brownien et réalité moléculaire. *Ann. Chim. Phys. (Paris)*, **19** (1909) 8:5–104. 1, 11

[5] P. HÄNGGI, J. LUCZKA, and P. TALKNER. World year of physics 2005: Focus on Brownian motion and diffusion in the 21st century. *New J. Phys.*, **7** (2005). 1

[6] W.R. SCHNEIDER and W. WYSS. Fractional diffusion and wave equations. *J. Math. Phys. (NY)*, **30** (1989) 1:134–144. 2, 87

[7] R. METZLER, E. BARKAI, and J. KLAFTER. Anomalous diffusion and relaxation close to thermal equilibrium: A fractional Fokker–Planck equation approach. *Phys. Rev. Lett.*, **82** (1999) 18:3563–3567. 88

[8] R. METZLER and J. KLAFTER. The random walk's guide to anomalous diffusion: A fractional dynamics approach. *Phys. Rep.*, **339** (2000):1–77. 2, 25, 84, 85, 87, 89, 159

[9] H. SCHER and E.W. MONTROLL. Anomalous transit-time dispersion in amorphous solids. *Phys. Rev. B*, **12** (1975) 6:2455–2477. 2, 25, 83, 85, 106

[10] K. HENZLER-WILDMAN and D. KERN. Dynamic personalities of proteins. *Nature*, **450** (2007): 964–972. 2, 4, 5

[11] J.F. ATKINS and R. GESTELAND. The 22nd amino acid. *Science*, **296** (2002) 5572:1409–1410. 3

[12] K. MARUYAMA, S. KIMURA, K. OHASHI, and Y. KUWANO. Connectin, an elastic protein of muscle. identification of "titin" with connectin. *J. Biochem.*, **89** (1981) 3:701–709. 3

[13] S. LABEIT, D.P. BARLOW, M. GAUTEL, T. GIBSON, J. HOLT, C.-L. HSIEH, U. FRANCKE, K. LEONARD, J. WARDALE, A. WHITING, and J. TRINICK. A regular pattern of two types of 100-residue motif in the sequence of titin. *Nature*, **345** (1990):273–276. 3

[14] J.C. KENDREW, G. BODO, H.M. DINTZIS, R.G. PARRISH, H. WYCKOFF, and D.C. PHILLIPS. A three-dimensional model of the myoglobin molecule obtained by X-ray analysis. *Nature*, **181** (1958):662–666. 4

[15] M.F. PERUTZ, M.G. ROSSMANN, ANN F. CULLIS, HILARY MUIRHEAD, GERORG WILL, and A.C.T. NORTH. Structure of haemoglobin: A three-dimensional fourier synthesis at 5.5å. resolution, obtained by x-ray analysis. *Nature*, **185** (1960):416–422. 4

[16] M. KARPLUS and G.A. PETSKO. Molecular dynamics simulations in biology. *Nature*, **347** (1990):631–639. 4

[17] M. KARPLUS and J.A. MCCAMMON. Molecular dynamics simulations of biomolecules. *Nat. Struc. Mol. Biol.*, **9** (2002):646–652.

[18] D. FRENKEL and B. SMIT. *Understanding Molecular Simulation: From Algorithms to Applications*, volume 1 of *Computationl Science*. Academic Press, San Diego, San Francisco, New York, Boston, London, Sydney, Tokyo, 2002. 4, 32, 33, 40, 167

[19] P.G. DEBRUNNER and H. FRAUENFELDER. Dynamics of proteins. *Ann. Rev. Phys. Chem.*, **33** (1982):283–299. 5

[20] A. ANSARI, J. BERENDZEN, S.F. BOWNE, H. FRAUENFELDER, I.E.T. IBEN, T.B. SAUKE, E. SHYAMSUNDER, and R.D. YOUNG. Protein states and proteinquakes. *Proc. Natl. Acad. Sci. USA*, **82** (1985):5000–5004. 5, 44, 83

[21] R. ELBER and M. KARPLUS. Multiple conformational states of proteins: a molecular dynamics analysis of myoglobin. *Science*, **235** (1987) 4786:318–321.

[22] H. FRAUENFELDER, S.G. SLIGAR, and P.G. WOLYNES. The energy landscapes and motions of proteins. *Science*, **254** (1991):1598–1603. 5, 44, 130

[23] R.H. AUSTIN, K. BEESON, L. EISENSTEIN, H. FRAUENFELDER, I.C. GUNSALUS, and V.P. MARSHALL. Activation energy spectrum of a biomolecule: Photodissociation of carbonmonoxy myoglobin at low temperatures. *Phys. Rev. Lett.*, **32** (1974) 8:403–405. 5, 6, 83, 130

[24] A. KITAO, S. HAYWARD, and N. GO. Energy landscape of a native protein: Jumping-among-minima model. *Proteins: Struct. Funct. Genet.*, **33** (1998):496. 5, 63, 83

[25] M. GOLDSTEIN. Viscous liquids and the glass transition: A potential energy barrier picture. *J. Chem. Phys.*, **51** (1969) 9:3728–3739. 5, 44

[26] C.A. ANGELL. Formation of glasses from liquids and biopolymers. *Science*, **267** (1995) 5206:1924–1935. 5, 44

[27] S.C. KOU and X. SUNNEY XIE. Generalized Langevin equation with fractional Gaussian noise: Subdiffusion within a single protein molecule. *Phys. Rev. Lett.*, **93** (2004) 18:180603. 5, 6, 82, 98, 130, 131

[28] W. MIN, G. LUO, B.J. CHERAYIL, S.C. KOU, and X. SUNNEY XIE. Observation of a power-law memory kernel for fluctuations within a single protein molecule. *Phys. Rev. Lett.*, **94** (2005) 19:198302. 6, 98, 130, 131

[29] P.W. FENIMORE, H. FRAUENFELDER, B.H. MCMAHON, and R.D. YOUNG. Bulk-solvent and hydration-shell fluctuations, similar to α- and β-fluctuations in glasses, control protein motions and functions. *Proc. Natl. Acad. Sci. USA*, **101** (2004) 40:14408–14413. 5, 83

[30] I.E.T. IBEN, D. BRAUNSTEIN, W. DOSTER, H. FRAUENFELDER, M.K. HONG, J.B. JOHNSON ANS S. LUCK, P. ORMOS, A. SCHULTE, P.J. STEINBACH, A.H. XIE, and R.D. YOUNG. Glassy behavior of a protein. *Phys. Rev. Lett.*, **62** (1989) 16:1916–1919. 5, 83, 130

[31] W. DOSTER, S. CUSACK, and W. PETRY. Dynamical transition of myoglobin revealed by inelastic neutron scattering. *Nature*, **337** (1989):754–756. 5, 55, 83, 133

[32] GIUSEPPE ZACCAI. How soft is a protein? a protein dynamics force constant measured by neutron scattering. *Science*, **288** (2000) 5471:1604–1607.

[33] G. CALISKAN, R.M. BRIBER, D. THIRUMALAI, V. GARCIA-SAKAI, S.A. WOODSON, and A.P. SOKOLOV. Dynamic transition in tRNA is solvent induced. *J. Am. Chem. Soc.*, **128** (2006) 1:32–33. 5

[34] S.-H. CHEN, L. LIU, E. FRATINI, P. BAGLIONI, A. FARAONE, and E. MAMONTOV. Observation of fragile-to-strong dynamic crossover in protein hydration water. *Proc. Natl. Acad. Sci. USA*, **103** (2006):9012–9016. 5

[35] D. RINGE and G.A. PETSKO. The 'glass transition' in protein dynamics: what it is, why it occurs, and how to exploit it. *Biophys. Chem.*, **105** (2003) 2-3:667–680. 5

[36] B.F. RASMUSSEN, A.M. STOCK, D. RINGE, and G.A. PETSKO. Crystalline ribonuclease A loses function below the dynamical transition at 220 k. *Nature*, **357** (1992):423–424. 5

[37] F. PARAK. Inter- and intramolecular motions in proteins. *Int. J. Quantum Chem.*, **42** (1992):1491–1498. 5

[38] S.G. COHEN, E.R. BAUMINGER, I. NOWIK, S. OFER, and J. YARIV. Dynamics of the iron-containing core in crystals of the iron-storage protein, ferritin, through mössbauer spectroscopy. *Phys. Rev. Lett.*, **46** (1981) 18:1244–1248.

[39] E.R. BAUMINGER, S.G. COHEN, I. NOWIK, S. OFER, and J. YARIV. Dynamics of heme iron in crystals of metmyoglobin and deoxymyoglobin. *Proc. Natl. Acad. Sci. USA*, **80** (1983) 3:736–740.

[40] H. LICHTENEGGER, W. DOSTER, T. KLEINERT, A. BIRK, B. SEPIOL, and G. VOGL. Heme-solvent coupling: A Mossbauer study of myoglobin in sucrose. *Biophys. J.*, **76** (1999) 1:414–422.

[41] F.G. PARAK. Physical aspects of protein dynamics. *Rep. Prog. Phys.*, **66** (2003) 2:103–129. 5

[42] A.L. TOURNIER, J. XU, and J.C. SMITH. Translational hydration water dynamics drives the protein glass transition. *Biophys. J.*, **85** (2003):1871–1875. 5, 40

[43] V. HELMS. Protein dynamics tightly connected to the dynamics of surrounding and internal water molecules. *Chem. Phys. Chem.*, **8** (2007): 23–33. 40

[44] N. SHENOGINA, P. KEBLINSKI, and S. GARDE. Strong frequency dependence of dynamical coupling between protein and water. *J. Chem. Phys.*, **129** (2008) 15:155105.

[45] L. ZHANG, L. WANG, Y.-T. KAO, W. QIU, Y. YANG, O. OKOBIAH, and D. ZHONG. Mapping hydration dynamics around a protein surface. *Proc. Natl. Acad. Sci. USA*, **104** (2007) 47:18461–18466. 5

[46] R.M. DANIEL, J.C. SMITH, M. FERRAND, S. HÉRY, R. DUNN, and J.L. FINNEY. Enzyme activity below the dynamical transition at 220 k. *Biophys. J.*, **75** (1998):2504–2507. 5

[47] R.V. DUNN, V. RÉAT, J. FINNEY, M. FERRAND, J.C. SMITH, and R.M. DANIEL. Enzyme activity and dynamics: xylanase activity in the absence of fast anharmonic dynamics. *Biochem. J.*, **346** (2000):355–358. 5

[48] Y. HE, P.I. KU, J.R. KNAB, J.Y. CHEN, and A.G. MARKELZ. Protein dynamical transition does not require protein structure. *Phys. Rev. Lett.*, **101** (2008) 17:178103. 5, 133

[49] S.C. KOU. Stochastic modeling in nanoscale biophysics: Subdiffusion within proteins. *Ann. Appl. Stat.*, **2** (2008) 2:501–535. 5

[50] W.G. GLÖCKLE and T.F. NONNENMACHER. A fractional calculus approach to self-similar protein dynamics. *Biophys. J.*, **68** (1995):46–53. 6, 130

[51] H. YANG and X. SUNNEY XIE. Probing single-molecule dynamics photon by photon. *J. Chem. Phys.*, **117** (2002) 24:10965–10979. 6, 82, 130

[52] H. YANG, G. LUO, P. KARNCHANAPHANURACH, T.-M. LOUIE, I. RECH, S. COVA, L. XUN, and X. SUNNEY XIE. Protein conformational dynamics probed by single-molecule electron transfer. *Science*, **302** (2003):262–266. 6, 82, 85, 08, 130

[53] G.R. KNELLER and K. HINSEN. Fractional Brownian dynamics in proteins. *J. Chem. Phys.*, **121** (2004) 20:10278–10283. 6, 99, 130, 131

[54] G.R. KNELLER. Quasielastic neutron scattering and relaxation processes in proteins: Analytical and simulation-based models. *Phys. Chem. Chem. Phys.*, **7** (2005) 13:2641–2655. 6, 98, 130, 131, 134

[55] GERO VOGL. *Wandern ohne Ziel*. Springer, Berlin, 2007. 9

[56] C. CERCIGNANI. *Ludwig Boltzmann: The Man Who Trusted Atoms*. Oxford University Press, Oxford, 2006. 9, 10, 11

[57] R. CLAUSIUS. Ueber die Art der Bewegung die wir Wärme nennen. *Pogg. Ann. d. Phys. u. Chem.*, **100** (1857):353–380. 10

[58] R. CLAUSIUS. Ueber verschiedene für die Anwendung bequeme Formen der Hauptgleichungen der mechanischen Wärmetheorie. *Pogg. Ann. d. Phys. u. Chem.*, **125** (1865):353–400. 10

[59] L. BOLTZMANN. Weitere Studien über das Wärmegleichgewicht unter Gasmolekülen. *Wien. Ber.*, **66** (1872):275–370. 10

[60] M. PLANCK. *Wissenschaftliche Selbstbiographie.* J.A. Barth, Leipzig; Göttingen, 1948. 10

[61] L. BOLTZMANN. Über die Eigenschaften monozyklischer und anderer damit verwandter Systeme. *Crelle's J.*, **98** (1885):68–94. 10

[62] M. VON SMOLUCHOWSKI. Zur kinetischen Theorie der Brownschen Molekularbewegung und der Suspensionen. *Ann. d. Phys.*, **21** (1906):756–780. 11

[63] K. PEARSON. The problem of the random walk. *Nature*, **72** (1905):294. 11, 23

[64] P. LANGEVIN. Sur la théorie du mouvement Brownien. *C. R. Acad. Sci. (Paris)*, **146** (1908):530–533. 11, 16

[65] L.E. REICHL. *A Modern Course in Statistical Physics.* Wiley, New York, 2nd edition, 1998. 12, 13, 14, 15, 24, 42

[66] H. DIETER ZEH. *The Physical Basis of The Direction of Time.* Springer, Berlin, 5th edition, 2007. 13

[67] R.P. FEYNMAN. *Six easy pieces : essentials of physics explained by its most brilliant teacher.* Perseus books, 1995. 13

[68] L. BOLTZMANN. Über die Natur der Gasmoleküle. *Wien. Ber.*, **74** (1876):553–560. 14

[69] R. ZWANZIG. *Nonequilibrium statistical mechanics.* Oxford University Press, 2001. 14, 16, 19, 24, 69, 75, 79, 83

[70] S. CHANDRASEKAR. Stochastic problems in physics and astronomy. *Rev. Mod. Phys.*, **15** (1943):1–89. 16, 19

[71] D. BEEMAN. Some multistep methods for use in molecular dynamics calculations. *J. Comp. Phys.*, **20** (1976) 2:130–139. 17, 21

[72] A. KITAO, F. HIRATA, and N. GO. The effects of solvent on the conformation and the collective motions of protein: normal mode analysis and molecular dynamics simulation of melittin in water and in vacuum. *Chem. Phys.*, **158** (1991):447–472. 21, 45

[73] S. HAYWARD, A. KITAO, F. HIRATA, and N. GO. Effect of solvent on collective motion in globular protein. *J. Mol. Biol.*, **234** (1993):1207. 21

[74] R. METZLER and J. KLAFTER. The restaurant at the end of the random walk. *J. Phys. A: Math. Gen.*, **37** (2004):R161–R208. 25, 85

[75] P.W.M. BLOM and M.C.J.M. VISSENBERG. Dispersive hole transport in poly(p-phenylene vinylene). *Phys. Rev. Lett.*, **80** (1998) 17:3819–3822. 26

[76] J. MASOLIVER, M. MONTERO, and G.H. WEISS. Continuous-time random-walk model for financial distributions. *Phys. Rev. E*, **67** (2003) 2: 021112. 26

[77] J. PERELLÓ, J. MASOLIVER, A. KASPRZAK, and R. KUTNER. Model for interevent times with long tails and multifractality in human communications: An application to financial trading. *Phys. Rev. E*, **78** (2008) 3:036108. 26

[78] I. GOLDING and E.C. COX. Physical nature of bacterial cytoplasm. *Phys. Rev. Lett.*, **96** (2006) 9:098102. 26, 85

[79] M. PLATANI, I. GOLDBERG, A.I. LAMOND, and J.R. SWEDLOW. Cajal body dynamics and association with chromatin are ATP-dependent. *Nat. Cell Biol.*, **4** (2002) 7:502–508. 85

[80] M. WEISS, H. HASHIMOTO, and T. NILSSON. Anomalous protein diffusion in living cells as seen by fluorescence correlation spectroscopy. *Biophys. J.*, **84** (2003) 6:4043–4052. 85

[81] M. WEISS, M. ELSNER, F. KARTBERG, and T. NILSSON. Anomalous subdiffusion is a measure for cytoplasmic crowding in living cells. *Biophys. J.*, **87** (2004) 5:3518–3524.

[82] G. GUIGAS and M. WEISS. Sampling the cell with anomalous diffusion: The discovery of slowness. *Biophys. J.*, **94** (2008) 1:90–94. 26, 85, 109

[83] I.M. TOLIĆ-NØRRELYKKE, E.-L. MUNTEANU, G. THON, L. ODDERSHEDE, and K. BERG-SØRENSEN. Anomalous diffusion in living yeast cells. *Phys. Rev. Lett.*, **93** (2004) 7:078102. 26, 85

[84] I.Y. WONG, M.L. GARDEL, D.R. REICHMAN, E.R. WEEKS, M.T. VALENTINE, A.R. BAUSCH, and D.A. WEITZ. Anomalous diffusion probes microstructure dynamics of entangled F-actin networks. *Phys. Rev. Lett.*, **92** (2004) 17:178101. 26, 85

[85] B. BERKOWITZ, A. CORTIS, M. DENTZ, and H. SCHER. Modeling non-Fickian transport in geological formations as a continuous time random walk. *Rev. Geophys.*, **44** (2006):RG2003. 26, 85, 106

[86] B. BERKOWITZ, S. EMMANUEL, and H. SCHER. Non-Fickian transport and multiple-rate mass transfer in porous media. *Water Resources Research*, **44** (2008):W03402. 26, 85, 106

[87] B.I. HENRY, T.A.M. LANGLANDS, and S.L. WEARNE. Fractional cable models for spiny neuronal dendrites. *Phys. Rev. Lett.*, **100** (2008) 12:128103. 26

[88] R. METZLER and J. KLAFTER. When translocation dynamics becomes anomalous. *Biophys. J.*, **85** (2003) 4:2776–2779. 26

[89] G.J. SCHUTZ, H. SCHINDLER, and T. SCHMIDT. Single-molecule microscopy on model membranes reveals anomalous diffusion. *Biophys. J.*, **73** (1997) 2:1073–1080. 26

[90] J. KLAFTER and I.M. SOKOLOV. Anomalous diffusion spreads its wings. *Physics today*, (2005):29. 26, 85

[91] E. JOOS, H.D. ZEH, C. KIEFER, D. GIULINI, J. KUPSCH, and I.-O. STAMATESCU. *Decoherence and the Appearance of a Classical World in Quantum Theory*. Springer, Berlin; Heidelberg, 2nd edition, 2003. 29

[92] R.F. SERVICE. Exploring the systems of life. *Science*, **284** (1999) 5411:80a–83. 29

[93] N. GOLDENFELD and L.P. KADANOFF. Simple lessons from complexity. *Science*, **284** (1999): 87–89. 29

[94] E.B. WALTON and K.J. VANVLIET. Equilibration of experimentally determined protein structures for molecular dynamics simulation. *Phys. Rev. E*, **74** (2006) 6:061901. 30

[95] M. BORN and R. OPPENHEIMER. Zur Quantentheorie der Molekeln. *Ann. Phys. (Leipzig)*, **84** (1927):457–484. 30

[96] B. HESS, H. BEKKER, H.J.C. BERENDSEN, and J.G.E.M. FRAAIJE. LINCS: A linear constraint solver for molecular simulations. *J. Comp. Chem.*, **18** (1997) 12:1463–1472. 31, 52, 155, 157

[97] D. VAN DER SPOEL, E. LINDAHL, and B. HESS. *GROMACS user manual: Version 3.3*. The GROMACS development team, 2006. URL http://www.gromacs.org/content/view/27/42/. ftp://ftp.gromacs.org/pub/manual/manual-3.3.pdf. 31, 37, 46, 51, 53, 157

[98] D. VAN DER SPOEL and H. J. C. BERENDSEN. *GROningen MAchine for Chemical Simulation*. Departement of Biophysical Chemistry, BIOSON Research Institute, Nijenborgh 4 NL-9717 AG Groningen, 1994. 31, 51

[99] W.F. VAN GUNSTEREN, S.R. BILLETER, A.A. EISING, P.H. HÜNENBERGER, P. KRÜGER, A.E. MARK, W.R.P. SCOTT, and I.G. TIRONI. *Biomolecular Simulation: The GROMOS96 Manual and User Guide*. Hochschulverlag AG an der ETH Zürich, 1996. 31, 34, 51

[100] L. VERLET. Computer "experiments" on classical fluids. I. thermodynamical properties of

Lennard-Jones molecules. *Phys. Rev.*, **159** (1967) 1:98. 32

[101] R.E. GILLILAN and K.R. WILSON. Shadowing, rare events, and rubber bands. a variational Verlet algorithm for molecular dynamics. *J. Chem. Phys.*, **97** (1992) 3:1757–1772. 33

[102] S. TOXVAERD. Hamiltonians for discrete dynamics. *Phys. Rev. E*, **50** (1994) 3:2271–2274. 33

[103] L. MEINHOLD. *Crystalline Protein Dynamics: A Simulation Analysis of Staphylococcal Nuclease*. PhD thesis, Physikalische Fakultät, Ruprecht-Karls-Universität, Heidelberg, 2005. URL http://www.ub.uni-heidelberg.de/archiv/5898. 35, 38, 49

[104] P. P. EWALD. Die berechnung optischer und elektrostatischer gitterpotentiale. *Ann. Phys. (Leipzig)*, **64** (1921) 3:253–287. 37

[105] T. DARDEN, D. YORK, and L. PEDERSEN. Particle mesh ewald: An $N-\log(N)$ method for Ewald sums in large systems. *J. Chem. Phys.*, **98** (1993) 12:10089. 40, 52

[106] U. ESSMANN, L. PERERA, M.L. BERKOWITZ, T. DARDEN, H. LEE, and L.G. PEDERSEN. A smooth particle mesh Ewald method. *J. Chem. Phys.*, **103** (1995) 19:8577–8593. 40

[107] W.H. PRESS, S.A. TEUKOLSKY, W.T. VETTERLING, and B.P. FLANNERY, editors. *Numerical recipes in C: The art of scientific computing*. Cambridge University Press, 1992. 40, 85, 86, 160

[108] A.N. DROZDOV, A. GROSSFIELD, and R.V. PAPPU. Role of solvent in determining conformational preferences of alanine dipeptide in water. *J. Am. Chem. Soc.*, **126** (2004) 8:2574–2581. 40

[109] P.G. DEBENEDETTI and H.E. STANLEY. Supercooled and glassy water. *Phys. Today*, **56** (2003) 6:40–46. 41

[110] H.E. STANLEY, P. KUMAR, G. FRANZESE, L. XU, Z. YAN, M.G. MAZZA, S.V. BULDYREV, S.-H. CHEN, and F. MALLAMACE. Liquid polyamorphism: Possible relation to the anomalous behavior of water. *Europ. Phys. J.: Spec. Topics*, **161** (2008):1–17.

[111] P. KUMAR, G. FRANZESE, and H.E. STANLEY. Dynamics and thermodynamics of water. *J. Phys.: Cond. Matt.*, **20** (2008) 24:244114 (12pp). 41

[112] H.J.C. BERENDSEN, J.R. GRIGERA, and T.P. STRAATSMA. The missing term in effective pair potentials. *J. Phys. Chem.*, **91** (1987):6269–6271. 41, 52

[113] W.C. STILL, A. TEMPCZYK, R.C. HAWLEY, and T. HENDRICKSON. Semianalytical treatment of solvation for molecular mechanics and dynamics. *J. Am. Chem. Soc.*, **112** (1990) 16:6127–6129. 41, 53

[114] D. BASHFORD and D.A. CASE. Generalized Born models of macromolecular solvation effects. *Annu. Rev. Phys. Chem.*, **51** (2000) 1:129–152. 41, 53

[115] B. MA and R. NUSSINOV. Explicit and implicit water simulations of a β-hairpin peptide. *Prot. Struc. Func. Gen.*, **37** (1999) 1:73–87. 41

[116] B.D. BURSULAYA and C.L. BROOKS. Comparative study of the folding free energy landscape of a three-stranded β-sheet protein with explicit and implicit solvent models. *J. Phys. Chem. B*, **104** (2000) 51:12378–12383. 41

[117] D. BROWN and J.H.R. CLARKE. A comparison of constant energy, constant temperature, and constant pressure ensembles in molecular dynamics simulations of atomic liquids. *Mol. Phys.*, **51** (1984):1243–1252. 42, 52

[118] D.J. EVANS and S. SARMAN. Equivalence of thermostatted nonlinear responses. *Phys. Rev. E*, **48** (1993) 1:65–70. 42

[119] H. FRAUENFELDER, F. PARAK, and R.D. YOUNG. Conformational substates in proteins. *Ann. Rev. Biophys. Biophys. Chem.*, **17** (1988):451–479. 44

[120] K. MORITSUGU and J.C. SMITH. Langevin model of the temperature and hydration dependence of protein vibrational dynamics. *J. Phys. Chem. B*, **109** (2005):12182–12194. 45

[121] K. PEARSON. On lines and planes of closest fit to systems of points in space. *Phil. Mag.*, **2** (1901) 6:559–572. 46

[122] A. AMADEI, A.B.M. LINSSEN, and H.J.C. BERENDSEN. Essential dynamics of proteins. *Proteins: Struct. Funct. Genet.*, **17** (1993) 4:412–425. 46

[123] D.M.F. VAN AALTEN, B.L. DE GROOT, J.B.C. FINDLAY, H.J.C. BERENDSEN, and A. AMADEI. A comparison of techniques for calculating protein essential dynamics. *J. Comp. Chem.*, **18** (1997) 2:169–181. 46, 48

[124] A.E. GARCÍA. Large-amplitude nonlinear motions in proteins. *Phys. Rev. Lett.*, **68** (1992) 17:2696–2699. 46

[125] R.M. LEVY, A.R. SRINIVASAN, W.K. OLSON, and J.A. MCCAMMON. Quasi-harmonic method for studying very low frequency modes in proteins. *Biopolymers*, **23** (1984) 6:1099–1112. 46

[126] I.T. JOLLIFFE. *Principal component analysis*. Springer, New York; Berlin; Heidelberg, 2nd edition, 2002. 46

[127] G.G. MAISURADZE and D.M. LEITNER. Free energy landscape of a biomolecule in dihedral principal component space: Sampling convergence and correspondence between structures and minima. *Proteins: Struct. Func. Bioinfo.*, **67** (2007) 3:569–578. 48

[128] Y.M. KOYAMA, T.J. KOBAYASHI, S. TOMODA, and H.R. UEDA. Perturbational formulation of principal component analysis in molecular dynamics simulation. *Phys. Rev. E*, **78** (2008) 4:046702. 48

[129] P.H. HÜNENBERGER, A.E. MARK, and W.F. VAN GUNSTEREN. Fluctuation and cross-correlation analysis of protein motions observed in nanosecond molecular dynamics simulations. *J. Mol. Biol.*, **252** (1995) 4:492–503. 48

[130] M. KARPLUS and T. ICHIYE. Comment on a "fluctuation and cross correlation analysis of protein motions observed in nanosecond molecular dynamics simulations". *J. Mol. Biol.*, **263** (1996) 2:120–122.

[131] R. ABSEHER and M. NILGES. Are there non-trivial dynamic cross-correlations in proteins? *J. Mol. Biol.*, **279** (1998) 4:911–920. 48

[132] L.S.D. CAVES, J.D. EVANSECK, and M. KARPLUS. Locally accessible conformations of proteins: Multiple molecular dynamics simulations of crambin. *Protein Sci.*, **7** (1998):649–666. 48

[133] M.A. BALSERA, W. WRIGGERS, Y. OONO, and K. SCHULTEN. Principal component analysis and long time protein dynamics. *J. Phys. Chem.*, **100** (1996) 7:2567–2572. 48, 49

[134] J. CAO and B.J. BERNE. On energy estimators in path integral monte carlo simulations: Dependence of accuracy on algorithm. *J. Chem. Phys.*, **91** (1989) 10:6359–6366.

[135] R.D. MOUNTAIN and D. THIRUMALAI. Measures of effective ergodic convergence in liquids. *J. Phys. Chem.*, **93** (1989) 19:6975–6979.

[136] B. HESS. Similarities between principal components of protein dynamics and random diffusion. *Phys. Rev. E*, **62** (2000) 6:8438–8448. 49, 130

[137] B. HESS. Convergence of sampling in protein simulations. *Phys. Rev. E*, **65** (2002) 3:031910. 49

[138] E. LYMAN and D.M. ZUCKERMAN. Ensemble-based convergence analysis of biomolecular trajectories. *Biophys. J.*, **91** (2006) 1:164–172. 49

[139] A. GROSSFIELD, S.E. FELLER, and M.C. PITMAN. Convergence of molecular dynamics simulations of membrane proteins. *Proteins: Struct. Funct. Genet.*, **67** (2007) 1:31–40. 49

[140] T. NEUSIUS, I. DAIDONE, I.M. SOKOLOV, and J.C. SMITH. Subdiffusion in peptides originates from the fractal-like structure of configuration space. *Phys. Rev. Lett.*, **100** (2008) 18:188103. 51, 75, 85, 90, 99

[141] I. DAIDONE, M.B. ULMSCHNEIDER, A. DI NOLA, A. AMADEI, and J.C. SMITH. Dehydration-driven solvent exposure of hydrophobic surfaces as a driving force in peptide folding. *Proc. Natl. Acad. Sci. USA*, **104** (2007) 39:15230–15235. 51, 52, 53

[142] D. QIU, P.S. SHENKIN, F.P. HOLLINGER, and W.C. STILL. The GB/SA continuum model for solvation. a fast analytical method for the calculation of approximate Born radii. *J. Phys. Chem. A*, **101** (1997) 16:3005–3014. 53

[143] J. ZHU, Y. SHI, and H. LIU. Parametrization of a generalized Born/solvent-accessible surface area model and applications to the simulation of protein dynamics. *J. Phys. Chem. B*, **106** (2002) 18:4844–4853. 53

[144] M. SCHAEFER, C. BARTELS, and M. KARPLUS. Solution conformations and thermodynamics of structured peptides: molecular dynamics simulation with an implicit solvation model. *J. Mol. Biol.*, **284** (1998) 3:835–848. 53

[145] A.L. TOURNIER and J.C. SMITH. Principal components of the protein dynamical transition. *Phys. Rev. Lett.*, **91** (2003) 20:208106. 54, 58, 59

[146] J.G. KIRKWOOD. Statistical mechanics of fluid mixtures. *J. Chem. Phys.*, **3** (1935) 5:300–313. 58

[147] T. NEUSIUS, I.M. SOKOLOV, and J.C. SMITH. Subdiffusion in time-averaged, confined random walks. *Phys. Rev. E*, **80** (2009) 1:011109. 75, 85, 93

[148] M. CERIOTTI, G. BUSSI, and M. PARRINELLO. Langevin equation with colored noise for constant-temperature molecular dynamics simulations. *Phys. Rev. Lett.*, **102** (2009) 2:020601. 79

[149] P.E. ROUSE. A theory of the linear viscoelastic properties of dilute solutions of coiling polymers. *J. Chem. Phys.*, **21** (1953) 7:1272–1280. 80, 157

[150] G. STROBL. *The Physics of Polymers*. Springer, Berlin; Heidelberg, 1997. 80, 157

[151] J. TANG and R.A. MARCUS. Chain dynamics and power-law distance fluctuations in single–molecule systems. *Phys. Rev. E*, **73** (2006) 2:022102. 81, 82, 131

[152] B.H. ZIMM. Dynamics of polymer molecules in dilute solution: Viscoelasticity, flow birefringence and dielectric loss. *J. Chem. Phys.*, **24** (1956) 2:269–278. 81

[153] K.S. KOSTOV and K.F. FREED. Long-time dynamics of met-enkephalin: Comparison of theory with Brownian dynamics simulations. *Biophys. J.*, **76** (1999) 1:149–163. 82

[154] P. DEBNATH, W. MIN, X.S. XIE, and B.J. CHERAYIL. Multiple time scale dynamics of distance fluctuations in a semiflexible polymer: A one-dimensional generalized langevin equation treatment. *J. Chem. Phys.*, **123** (2005):204903. 82

[155] R. GRANEK and J. KLAFTER. Fractons in proteins: Can they lead to anomalously decaying time autocorrelations? *Phys. Rev. Lett.*, **95** (2005):098106. 82, 131

[156] S. REUVENI, R. GRANEK, and J. KLAFTER. Proteins: Coexistence of stability and flexibility. *Phys. Rev. Lett.*, **100** (2008) 20:208101. 82, 131

[157] H. NEUWEILLER, M. LÖLLMANN, S. DOOSE, and M. SAUER. Dynamics of unfolded polypeptide chains in crowded environment studied by fluorescence correlation spectroscopy. *J. Mol. Biol.*, **365** (2007) 3:856–869. 82

[158] J. TANG and S.-H. LIN. Distance versus energy fluctuations and electron transfer in single protein molecules. *Phys. Rev. E*, **73** (2006) 6: 061108. 82, 157

[159] R.M. VENABLE and R.W. PASTOR. Frictional models for stochastic simulations of proteins. *Biopolymers*, **27** (1988) 6:1001–1014. 82

[160] C. MONTHUS and J.-P. BOUCHAUD. Models of traps and glass phenomenology. *J. Phys. A: Math. Gen.*, **29** (1996):3847. 83

[161] S. BUROV and E. BARKAI. Occupation time statistics in the quenched trap model. *Phys. Rev. Lett.*, **98** (2007) 25:250601. 83

[162] J.-P. BOUCHAUD and A. GEORGES. Anomalous diffusion in disordered media: Statistical mechanisms, models and physical applications. *Phys. Rep.*, **195** (1990) 4-5:127–293. 83, 107

[163] Q. GU, E. A. SCHIFF, S. GREBNER, F. WANG, and R. SCHWARZ. Non-gaussian transport measurements and the Einstein relation in amorphous silicon. *Phys. Rev. Lett.*, **76** (1996) 17: 3196–3199. 85

[164] J.W. HAUS and K.W. KEHR. Diffusion in regular and disordered lattices. *Phys. Rep.*, **150** (1987) 5/6:263–406. 85, 110

[165] Y. HE, S. BUROV, R. METZLER, and E. BARKAI. Random time-scale invariant diffusion and transport coefficients. *Phys. Rev. Lett.*, **101** (2008) 5:058101. 85, 90, 97, 98

[166] KEN RITCHIE, XIAO-YUAN SHAN, JUNKO KONDO, KOKORO IWASAWA, TAKAHIRO FUJIWARA, and AKIHIRO KUSUMI. Detection of non-Brownian diffusion in the cell membrane in single molecule tracking. *Biophys. J.*, **88** (2005) 3:2266–2277. 85

[167] S. CONDAMIN, V. TEJEDOR, R. VOITURIEZ, O. BÉNICHOU, and J. KLAFTER. Probing microscopic origins of confined subdiffusion by first-passage observables. *Proc. Natl. Acad. Sci. USA*, **105** (2008) 15:5675–5680. 85

[168] M. WACHSMUTH, W. WALDECK, and J. LANGOWSKI. Anomalous diffusion of fluorescent probes inside living cell nuclei investigated by spatially-resolved fluorescence correlation spectroscopy. *J. Mol. Biol.*, **298** (2000) 4:677–689. 85

[169] S. YAMADA, D. WIRTZ, and S.C. KUO. Mechanics of living cells measured by laser tracking microrheology. *Biophys. J.*, **78** (2000) 4:1736–1747. 85

[170] M.A. DEVERALL, E. GINDL, E.-K. SINNER, H. BESIR, J. RUEHE, M.J. SAXTON, and C.A. NAUMANN. Membrane lateral mobility obstructed by polymer-tethered lipids studied at the single molecule level. *Biophys. J.*, **88** (2005) 3:1875–1886. 85

[171] V. NECHYPORUK-ZLOY, P. DIETERICH, H. OBERLEITHNER, C. STOCK, and A. SCHWAB. Dynamics of single potassium channel proteins in the plasma membrane of migrating cells. *Am. J. Physiol. Cell. Physiol.*, **294** (2008) 4: C1096–1102.

[172] Y.M. UMEMURA, M. VRLJIC, S.Y. NISHIMURA, T.K. FUJIWARA, K.G.N. SUZUKI, and A. KUSUMI. Both MHC class II and its GPI-anchored form undergo hop diffusion as observed by single-molecule tracking. *Biophys. J.*, **95** (2008) 1:435–450.

[173] S. WIESER, G.J. SCHÜTZ, M.E. COOPER, and H. STOCKINGER. Single molecule diffusion analysis on cellular nanotubules: Implications on plasma membrane structure below the diffraction limit. *Appl. Phys. Lett.*, **91** (2007) 23: 233901. 85

[174] J. SUNG and R.J. SILBEY. Exact dynamics of a continuous time random walker in the presence of a boundary: Beyond the intuitive boundary condition approach. *Phys. Rev. Lett.*, **91** (2003) 16:160601. 85

[175] M.A. LOMHOLT, I.M. ZAID, and R. METZLER. Subdiffusion and weak ergodicity breaking in the presence of a reactive boundary. *Phys. Rev. Lett.*, **98** (2007) 20:200603.

[176] M.L. HENLE, B. DIDONNA, C.D. SANTANGELO, and A. GOPINATHAN. Diffusion and binding of finite-size particles in confined geometries. *Phys. Rev. E*, **78** (2008) 3:031118. 85

[177] B. LIN, M. MERON, B. CUI, S.A. RICE, and H. DIAMANT. From random walk to single-file diffusion. *Phys. Rev. Lett.*, **94** (2005) 21: 216001. 85

[178] Q.-H. WEI, C. BECHINGER, and P. LEIDERER. Single-file diffusion of colloids in one-dimensional channels. *Science*, **287** (2000) 5453:625–627.

[179] A. RUSTOM, R. SAFFRICH, I. MARKOVIC, P. WALTHER, and H.-H. GERDES. Nanotubular highways for intercellular organelle transport. *Science*, **303** (2004) 5660:1007–1010.

[180] D. ODDE. Diffusion inside microtubules. *Eur. Biophys. J.*, **27** (1998) 5:514–520.

[181] H. NIKAIDO. Molecular basis of bacterial outer membrane permeability revisited. *Microbiol. Mol. Biol. Rev.*, **67** (2003) 4:593–656.

[182] T. CHOU and D. LOHSE. Entropy-driven pumping in zeolites and biological channels. *Phys. Rev. Lett.*, **82** (1999) 17:3552–3555. 85

[183] G. LUO, I. ANDRICIOAEI, X. SUNNEY XIE, and M. KARPLUS. Dynamic distance disorder in proteins is caused by trapping. *J. Phys. Chem. B*, **110** (2006) 19:9363. 85, 98, 130, 131

[184] G. BEL and E. BARKAI. Weak ergodicity breaking in the continuous-time random walk. *Phys. Rev. Lett.*, **94** (2005) 24:240602. 85, 88, 90, 91

[185] G. BEL and E. BARKAI. Random walk to a nonergodic equilibrium concept. *Phys. Rev. E*, **73** (2006) 1:016125. 91

[186] G. BEL and E. BARKAI. Weak ergodicity breaking with deterministic dynamics. *Europhys. Lett.*, **74** (2006) 1:15–21. 88

[187] A. REBENSHTOK and E. BARKAI. Distribution of time-averaged observables for weak ergodicity breaking. *Phys. Rev. Lett.*, **99** (2007) 21: 210601. 91

[188] A. LUBELSKI, I.M. SOKOLOV, and J. KLAFTER. Nonergodicity mimics inhomogeneity in single particle tracking. *Phys. Rev. Lett.*, **100** (2008) 25:250602. 85, 90

[189] E. BARKAI and Y.-C. CHENG. Aging continuous time random walks. *J. Chem. Phys.*, **118** (2003) 14:6167. 90, 93

[190] I.M. SOKOLOV, A. BLUMEN, and J. KLAFTER. Dynamics of annealed systems under external fields: CTRW and the fractional Fokker-Planck equations. *Europhys. Lett.*, **56** (2001) 2:175–180. 90

[191] I.M. SOKOLOV, A. BLUMEN, and J. KLAFTER. Linear response in complex systems: CTRW and the fractional Fokker-Planck equations. *Physica A*, **302** (2001) 1-4:268 – 278. 90

[192] N.G. VAN KAMPEN. *Stochastic processes in physics and chemistry*. Elsevier, Amsterdam, 5th edition, 2004. 91, 121

[193] S.H. STROGATZ. Exploring complex networks. *Nature*, **410** (2001) 6825:268–276. 102

[194] R. ALBERT and A.-L. BARABÁSI. Statistical mechanics of complex networks. *Rev. Mod. Phys.*, **74** (2002) 1:47–97. 102

[195] C.L. BROOKS III, J.N. ONUCHIC, and D.J. WALES. Taking a walk on a landscape. *Science*, **293** (2001) 5530:612–613. 103, 114, 130

[196] S. HAVLIN and D. BEN-AVRAHAM. Diffusion in disordered media. *Adv. Phys.*, **36** (1987):695–798. 107, 112, 113

[197] Y. GEFEN, A. AHARONY, B.B. MANDELBROT, and S. KIRKPATRICK. Solvable fractal family, and its possible relation to the backbone at percolation. *Phys. Rev. Lett.*, **47** (1981) 25: 1771–1774. 110

[198] Y. GEFEN, A. AHARONY, and S. ALEXANDER. Anomalous diffusion on percolating clusters. *Phys. Rev. Lett.*, **50** (1983) 1:77–80. 110

[199] S. HAVLIN, D. BEN-AVRAHAM, and D. MOVSHOVITZ. Percolation on fractal lattices. *Phys. Rev. Lett.*, **51** (1983) 26: 2347–2350. 110

[200] D.J. WALES. *Energy landscapes*. Cambridge University Press, Cambridge, 2003. 113

[201] F.H. STILLINGER and T.A. WEBER. Hidden structure in liquids. *Phys. Rev. A*, **25** (1982) 2:978–989. 113

[202] S. BÜCHNER and A. HEUER. Potential energy landscape of a model glass former: Thermodynamics, anharmonicities, and finite size effects. *Phys. Rev. E*, **60** (1999) 6:6507–6518. 114

[203] S. BÜCHNER and A. HEUER. Metastable states as a key to the dynamics of supercooled liquids. *Phys. Rev. Lett.*, **84** (2000):2168. 114

[204] F. NOÉ, D. KRACHTUS, J.C. SMITH, and S. FISCHER. Transition networks for the comprehensive characterization of complex conformational change in proteins. *J. Chem. Theory Comp.*, **2** (2006) 3:840–857. 114, 133

[205] S. AUER, M.A. MILLER, S.V. KRIVOV, C.M. DOBSON, M. KARPLUS, and M. VENDRUSCOLO. Importance of metastable states in the free energy landscapes of polypeptide chains. *Phys. Rev. Lett.*, **99** (2007) 17:178104. 114

[206] F. NOÉ, I. HORENKO, C. SCHÜTTE, and J.C. SMITH. Hierarchical analysis of conformational dynamics in biomolecules: Transition networks of metastable states. *J. Chem. Phys.*, **126** (2007) 15:155102. 114, 133

[207] J.D. CHODERA, N. SINGHAL, V.S. PANDE, K.A. DILL, and W.C. SWOPE. Automatic discovery of metastable states for the construction of Markov models of macromolecular conformational dynamics. *J. Chem. Phys.*, **126** (2007) 15:155101. 114

[208] M.A. MILLER and D.J. WALES. Energy landscape of a model protein. *J. Chem. Phys.*, **111** (1999) 14:6610–6616. 114

[209] J.P.K. DOYE. Network topology of a potential energy landscape: A static scale-free network. *Phys. Rev. Lett.*, **88** (2002) 23:238701. 114, 133

[210] D. GFELLER, D. MORTON DE LACHAPELLE, P. DE LOS RIOS, G. CALDARELLI, and F. RAO. Uncovering the topology of configuration space networks. *Phys. Rev. E*, **76** (2007):026113. 114

[211] J.P.K. DOYE and C.P. MASSEN. Characterizing the network topology of the energy landscapes of atomic clusters. *J. Chem. Phys.*, **122** (2005) 8:084105. 114

[212] J.D. BRYNGELSON, J.N. ONUCHIC, N.D. SOCCI, and P.G. WOLYNES. Funnels, pathways and the energy landscape of protein folding: A synthesis. *Proteins: Struct. Funct. Genet.*, **21** (1995): 167. 114

[213] D. GFELLER, P. DE LOS RIOS, A. CAFLISCH, and F. RAO. Complex network analysis of free-energy landscapes. *Proc. Natl. Acad. Sci. USA*, **104** (2007) 6:1817–1822. 114, 133

[214] S.V. KRIVOV and M. KARPLUS. Hidden complexity of free energy surfaces for peptide (protein) folding. *Proc. Natl. Acad. Sci. USA*, **101** (2004) 41:14766–14770. 114

[215] S.V. KRIVOV and M. KARPLUS. One-dimensional free-energy profiles of complex systems: Progress variables that preserve the barriers. *J. Phys. Chem. B*, **110** (2006) 25:12689–12698. 114

[216] S.V. KRIVOV and M. KARPLUS. Diffusive reaction dynamics on invariant free energy profiles. *Proc. Natl. Acad. Sci. USA*, **105** (2008) 37:13841–13846. 114

[217] C.-B. LI, H. YANG, and T. KOMATSUZAKI. Multiscale complex network of protein conformational fluctuations in single-molecule time series. *Proc. Natl. Acad. Sci. USA*, **105** (2008) 2:536–541. 114, 133

[218] N. HORI, G. CHIKENJI, R.S. BERRY, and S. TAKADA. Folding energy landscape and network dynamics of small globular proteins. *Proc. Natl. Acad. Sci. USA*, **106** (2009) 1:73–78. 114, 133

[219] C.B. SCHULZKY. *Anomalous diffusion and random walks on fractals*. PhD thesis, Fakultät für Naturwissenschaften, TU Chemnitz, 2000. URL http://archiv.tu-chemnitz.de/pub/. 125

[220] M.T. BARLOW and E.A. PERKINS. Brownian motion on the Sierpinski gasket. *Prob. Theo. Rel. Fields*, **79** (1988) 4:543–623. 125

[221] H. FRAUENFELDER, G.A. PETSKO, and D. TSERNOGLOU. Temperature-dependent X-ray diffraction as a probe of protein structural dynamics. *Nature*, **280** (1979) 5723: 558–563. 130

[222] F.H. STILLINGER. A topographic view of supercooled liquids and glass formation. *Science*, **267** (1995) 5206:1935–1939. 130

[223] FRANK NOÉ. Probability distributions of molecular observables computed from Markov models. *J. Chem. Phys.*, **128** (2008) 24:244103. 133

[224] C. NEUMANN. *Untersuchungen über das logarithmische und Newton'sche Potential*. B.G. Teubner, Leipzig, 1877. 157

Abbreviations and Symbols

Abbreviations

ACF	auto-correlation function
ACTRW	aging CTRW
CACF	coordinate ACF
CTRW	continuous time random walk
eSPC	extended simple point charge
FDE	fractional diffusion equation
FFT	fast Fourier transform
FPTD	first passage time distribution
GB/SA	generalized Born/surface area
GLE	generalized Langevin equation
JLD	jump length distribution
LINCS	linear constraints solver
MD	molecular dynamics
MLF	Mittag-Leffler function
MSD	mean squared displacement
NMA	normal mode analysis
NMR	nuclear magnetic resonance
PC	principle component
PCA	principle component analysis
PMF	potential of mean force
SPC	simple point charge
VACF	velocity ACF
WTD	waiting time distribution

Greek symbols

α	MSD exponent, Eq. (2.37)
Δt	simulation time step or random walk time step
Δ_k	anharmonicity degree, Eq. (4.10)
Γ	phase space
$\mathrm{d}\Gamma$	integration over phase space
$\Gamma(\alpha)$	Gamma function, Eq. (D.38)
$\Gamma(\alpha, x)$	incomplete Gamma function Eq. (D.42)
Γ_n	Sierpiński triangle of order n
Γ_∞	Sierpiński gasket
γ	Langevin friction, Eq. (2.11)
δ_{ij}	Kronecker's delta
$\delta(t)$	Dirac's delta function
η_k	participation ratio, Eq. (4.6)
ϑ	lag time of transition matrix, Eq. (5.63)
κ_α	prefactor, $= \sin(\pi\alpha)/\pi\alpha$
Λ_k	convergence factor, Eq. (4.12)
$\mu(t)$	memory function, Eq. (5.10)
$\xi, \boldsymbol{\xi}$	Langevin random force, Eq. (2.11)
$\chi_n(t_2, t_1)$	probability of n jumps in $[t_1, t_2]$, Eq. (5.27)
ρ	phase space density
τ_0	typical time scale in WTD Eq. (5.20)
τ_R	Rouse time, Eq. (5.17)
$\varphi(x)$	JLD

Latin symbols

$\mathbf{A}^T$	transposed of matrix $\mathbf{A}$

Abbreviations and Symbols

$\mathbf{C}$	covariance matrix, Eq. (3.47)	$\mathbf{S}(\vartheta)$	transition matrix on time scale ϑ, Eq. (5.63)
$C_v(t)$	VACF, Eq. (2.26)	$\tilde{\mathbf{S}}(\vartheta)$	approximate transition matrix
$C_x(t)$	CACF, Eq. (2.25)	T	length of observation/length of simulation
D	diffusion constant, Eq. (2.33)		
$_0\mathcal{D}_t^{1-\alpha}$	Riemann-Liouville operator, Eq. (5.23)	$\mathcal{T}$	temperature
d_f	Hausdorff or fractal dimension, Eq. (5.64)	$\mathcal{T}_g$	glass transition temp.
		t_c	critical time in confined CTRW, Eq. (5.55)
d_s	spectral dimension, Eq. (5.73)		
d_t	topological dimension	t_p	saturation time
d_w	diffusion dimension, Eq. (5.68)	t_s	start of observation in ACTRW
$E_\alpha(x)$	MLF, Eq. (5.24)	V	volume
$\mathcal{E}$	set of edges in network	$V(\mathbf{r})$	energy potential
$F(N,V,\mathcal{T})$	free energy, Eq. (4.3)	$\mathbf{v}$	velocity vector in configuration space
$\mathbf{f}$	force vector in configuration space, Eq. (3.1)		
		$W(x,t)$	probability distribution in diffusion equation
k_B	Boltzmann constant		
K_α	generalized diffusion constant, Eq. (5.22)	$w(t)$	WTD, Eq. (5.20)
		$w_1(t,t_s)$	initial WTD
$\mathbf{L}$	Liouville operator, Eq. (5.2)	$\mathbf{x}$	mass-weighted configuration vector
$\mathcal{L}$	Laplace transform, Eq. (D.53)		
$L^2(V)$	space of square integrable functions over V	$\mathbf{Z}$	counting matrix, Eq. (5.57)
		$\mathcal{Z}$	partition function
$\mathbf{M}$	mass matrix		
N	number of particles	**Other symbols**	
$\mathcal{N}$	set of vertices in network		
$n_{\mu,\sigma}$	Gaussian distribution with mean value μ and standart deviation σ	$\langle A\|B\rangle$	L^2-scalar product
		$\langle\cdot\rangle_\tau$	time average
$\mathcal{O}(\|x\|)$	Landau symbol	$\langle\cdot\rangle_{\tau,T}$	time average over $\tau\in[0,T]$
$\mathbf{p}^s$	stationary probability distribution	$\langle\cdot\rangle_{ens}$	ensemble average
		$\langle\cdot\rangle_{ens,0}$	CTRW: ensemble average, time relative to first jump
$\mathbf{q}$	PC vector		
R_k	transport coefficients, Eq. (5.92)	$\langle\cdot\rangle_{ens,t_s}$	CTRW: ensemble average, time relative to $t_s>0$
$\mathbf{r}$	position vector in configuration space		
		$\langle\cdot\rangle_t$	CTRW: combined ensemble and time average, also $\langle\langle\cdot\rangle\rangle$
$\mathbf{S}$	transition matrix, Eq. (5.63)		
		$\langle\Delta x^2(t)\rangle$	MSD

INDEX

ACF, 20, 21, 157
anharmonicity degree, 64
anharmonicity factor, 63
anomalous diffusion, 25

Bernoulli, Daniel, 10
Berzelius, Jöns Jakob, 2
Birkhoff, George David, 11
Boltzmann, Ludwig, 1, 10, 15
Born-Oppenheimer approximation, 30
Brown, Robert, 1
Brownian motion, 1, 2, 11, 16, 17, 49, 75

CACF, 20–22
central limit theorem, 17
Clausius, Rudolf, 10
configuration space, 43
convergence, 48, 68, 70
convergence factor, 72
counting matrix, 104
covariance matrix, 47, 55, 62
CTRW, 2, 83
    confined, 95

detailed balance, 105
diffusion constant, 24, 41

diffusion equation, 24, 153
dimension
    diffusion, 111
    fractal, 107
    spectral, 113
    topological, 107

Einstein relation, 24
Einstein, Albert, 1, 11, 16
energy landscape, 5, 42
ensemble average, 13, 87
equipartition theorem, 14
ergodicity, 10, 15, 16, 19–21, 48, 88, 90, 112
eSPC, 41, 74
Ewald, Paul Peter, 37

FDE, 87, 159, 161
Feynman, Richard, 13
Fick, Adolf, 1, 9
first law of thermodynamics, 9
fluctuation-dissipation theorem, 18
force field, 29, 34
Fourier decomposition, 9, 24, 88, 153
Fourier transform, 162
Fourier, Joseph, 9

FPTD, 69
fractal dynamics, 110
fractal geometry, 107
fractional derivative, 160
free energy, 56

Gamma function, 159
GB/SA, 41
generalized Born, 41
generalized Langevin equation, 78
Gibbs, Josiah Willard, 1
glass, 5, 44, 83
glass transition, 5, 54, 55, 59

Hausdorff dimension, 107
Hausdorff, Felix, 107
Helmholtz, Hermann von, 1, 9

isokinetic thermostat, 52

JLD, 83
Joule, James Prescott, 9

kinetic theory, 10
Kramer's escape theory, 69, 83

Langevin equation, 16, 19, 75, 78
Langevin thermostat, 41
Langevin, Paul, 11
Laplace transform, 162
Leapfrog algorithm, 34
LINCS algorithm, 52, 155
Liouville equation, 76, 78
Liouville theorem, 12

Mach, Ernst, 10

Mandelbrot, Benoît, 107
Markov property, 76, 78, 112
mass matrix, 12
mass-weighted coordinates, 44, 55
Maxwell, James Clerk, 10
Maxwell-Boltzmann distribution, 14, 19, 20, 31, 43
Mayer, Julius Robert von, 1, 9
MD simulation, 29
mean squared displacement, 19
Mittag-Leffler function, 88, 163
molecular disorder, 14
MSD, 19, 24
MSD exponent, 25

Neumann, Johann von, 11
neutron scattering, 4–6, 55
Newton's second law, 12
NMA, 44, 45
normal modes, 45

operational time, 90
Ostwald, Wilhelm, 10

participation number, 61, 62
participation ratio, 61
particle mesh Ewald method, 37, 40, 52
PCA, 46, 49, 62
Pearson, Karl, 11, 23, 46
peptide, 2, 3
Perrin, Jean, 1, 10, 11
phase space, 11
Planck, Max, 10

PMF, 56
Poisson equation, 38–40
potential of mean force, 56
principal components, 47
protein, 2, 3
protein folding, 4

random force, 16, 17
random walk, 2, 23
reaction coordinate, 56, 61
Riemann-Liouville, 87
Riemann-Liouville operator, 161
Rouse chain, 79, 157

second law of thermodynamics, 10, 14
secondary structure, 4, 5, 31, 52
Sierpiński, Wacław, 108
Sierpiński gasket, 108
Sierpiński triangle, 108
Smoluchowski, Marian von, 11
subdiffusion, 5, 6, 25
superdiffusion, 25

thermostat
    isokinetic, 41
    Langevin, 41
time average, 15, 90
trajectory, 13, 30
transition matrix, 105
transition network, 103
transport coefficient, 124
trap model, 3, 83, 85

VACF, 21, 22
Verlet algorithm, 32–34

water, 5, 40, 51, 66
    explicit, 41, 53
    implicit, 41, 53
white noise, 17, 65
WTD, 84, 86
WTD exponent, 84

Zermelo, Ernst, 10
Zwanzig's projection formalism, 75

> Ainsi il s'en s[u]it, que $d^{1:2}x$ sera egal à $x\sqrt[2]{dx:x}$. Il y a de l'apperance, qu'on tirera un jour des consequences bien utils de ces paradoxes, car il n'y a gueres des paradoxes sans utilité.
>
> G. W. F. LEIBNIZ

APPENDIX

A Diffusion equation solved by Fourier decomposition

The diffusion equation can be solved using Fourier decomposition. The principal idea is the separation of the time variable t and the coordinate x in the probability $W(x,t)$. The eigenfunctions of the Laplace operator in the diffusion equation, Eq. (2.33),

$$\frac{\partial}{\partial t}W(x,t) = D\frac{\partial^2}{\partial x^2}W(x,t) \tag{A.1}$$

can be expressed as imaginary exponentials, likewise the time dependence is given by a real exponential. Thus, the probability $W(x,t)$ is obtained as

$$W(x,t) = \int_{-\infty}^{\infty} a(k)e^{ikx}e^{-Dk^2 t}dk. \tag{A.2}$$

The initial condition and the boundary condition determine the Fourier coefficients $a(k)$. In the case of free, unbounded diffusion, the initial condition $W_0(x) = \delta(x)$ corresponds to $a(k) = (2\pi)^{-1}$. Then, Eq. (A.2) yields the probability distribution $W(x,t)$ of Eq. (2.34), which solves the diffusion equation. The MSD is obtained from the solution by integration. The result of the integrations is given in Eq. (2.35).

Diffusion with finite volume

In the presence of a reflecting boundary at $x = 0$ and at $x = L$, the probability distribution, $W(x,t)$ is confined to the coordinate interval $x \in [0, L]$. At the boundaries, the Dirichlet condition is satisfied

$$\frac{\partial}{\partial x}W(x,t)\bigg|_{x=0} = \frac{\partial}{\partial x}W(x,t)\bigg|_{x=L} = 0. \quad (A.3)$$

The boundary condition is fulfilled by the functions $\cos(\pi k x/L)$, with $k \in \mathbb{N}$. As a consequence, the integral in Eq. (A.2) is replaced by a discrete sum containing only the above cosine terms. Correspondingly, the time dependence has to be modified. We impose the initial condition $W_0(x) = \delta(x - x_0)$. The probability reads

$$W(x,t;x_0) = \frac{1}{L} + \frac{2}{L}\sum_{k=1}^{\infty} a_k(x_0) \cos\frac{\pi k x}{L} \exp\left(-\frac{4\pi^2 k^2}{L^2}Dx\right). \quad (A.4)$$

Inserting the probability distribution, $W(x,t;x_0)$ in Eq. (2.35), the MSD of the finite volume diffusion follows as

$$\langle \Delta x^2(t) \rangle = \frac{L^2}{6}\left[1 - \frac{96}{\pi^4}\sum_{k=0}^{\infty}\frac{1}{(2k+1)^4}\exp\left(-\frac{(2k+1)^2\pi^2}{L^2}Dt\right)\right] \quad (A.5)$$

$$= \frac{L^2}{6}\left[1 - \sum_{k=0}^{\infty}\theta_k g_k^t\right], \quad (A.6)$$

where $\theta_k = 96/[\pi(2k+1)]^4$ and

$$g_k = \exp\left(-\frac{(2k+1)^2\pi^2}{L^2}D\right). \quad (A.7)$$

In the main text, the following approximation is used for simplicity

$$\langle \Delta x^2(t) \rangle = \frac{L^2}{6}[1 - g^t], \quad (A.8)$$

with $g = \exp(-12D/L^2)$. Eq. (A.8) reproduces the behavior of the full solution for small times $t \to 0$ and it reaches the correct saturation level in the limit $t \to \infty$. However, it fails to reproduce the slowest relaxation time by a factor $\pi^4/12 \approx 0.8225$, i.e., the relaxation time is underestimated by 18%. Note that all calculations performed in Sec. 5.3 using this approximation can also be analogously carried out with the exact solution to Eq. (5.27).

B Constrained Verlet algorithm – LINCS

The fastest oscillations in an MD simulation are the bond vibrations which impose an upper bound to the maximal integration time step Δt. In order to increase the time step, the bond oscillations can be removed from the simulation. It turns out that simulations with fixed bond lengths are a more accurate representation of the physical behavior [96].

To remove bond oscillations, the bond lengths are holonomically constrained to a fixed length. Holonomic constraints can be introduced in the Verlet algorithm with the Linear Constraint Solver (LINCS) algorithm [96], which is briefly reviewed in the following.

The positions of a system with N particles are given by the $3N$ vector $\boldsymbol{r}(t)$, the equations of motion are yield from Newton's second principle, Eq. (3.2). The forces are assumed to be conservative forces, *i.e.*, they can be expressed as $\boldsymbol{f} = -\partial V/\partial \boldsymbol{r}$. Eq. (3.2) is a set of $3N$ differential equations. The holonomic, *i.e.* velocity independent, constraints are given by the equations

$$g_i(\boldsymbol{r}) = 0 \qquad i = 1, ..., K, \tag{B.9}$$

where the constraints are assumed to not have an explicit time dependence. Here, to fix the bond lengths, the functions $g_i(\boldsymbol{r})$ read

$$g_i(\boldsymbol{r}) = |\boldsymbol{r}^{i_1} - \boldsymbol{r}^{i_2}| - d_i \qquad i = 1, ..., K, \tag{B.10}$$

where d_i is the length of bond i between the atoms i_1 and i_2. The vectors with superscript denote the positions of the corresponding atoms.

Lagrange's first method allows the constraints to be included; the potential $V(\boldsymbol{r})$ is modified such that it contains the constraints multiplied with the Lagrange multipliers, $\lambda_i(t)$,

$$-\mathbf{M}\frac{\mathrm{d}^2 \boldsymbol{r}}{\mathrm{d}t^2} = \frac{\partial}{\partial \boldsymbol{r}}\left(V - \sum_i \lambda_i g_i\right). \tag{B.11}$$

With the notation $B_{hi} = \partial g_h/\partial r_i$, a $K \times 3N$ matrix $\mathbf{B}$ is defined. Using $\mathbf{B}$, Eq. (B.11) can be expressed as matrix equation

$$\frac{\mathrm{d}^2 \boldsymbol{r}}{\mathrm{d}t^2} = \mathbf{M}^{-1}\mathbf{B}^T \boldsymbol{\lambda} + \mathbf{M}^{-1}\boldsymbol{f}. \tag{B.12}$$

In the next step, the Lagrange multipliers, $\lambda_i(t)$, are determined from Eq. (B.12). Solving K of the $3N$ differential equations of Eq. (B.12) reduces the problem to $3N - K$ dimensions, in which the constraints of Eq. (B.9) are always respected.

The second time derivative of the constraints, Eq. (B.9), yields

$$-\mathbf{B}\frac{d^2 r}{dt^2} = \frac{d\mathbf{B}}{dt}\frac{dr}{dt}. \tag{B.13}$$

After multiplying Eq. (B.12) with $\mathbf{B}$ and using Eq. (B.13), one has

$$\mathbf{BM}^{-1}\mathbf{B}^T \boldsymbol{\lambda} = -\left(\frac{d\mathbf{B}}{dt}\frac{dr}{dt} + \mathbf{BM}^{-1}\boldsymbol{f}\right). \tag{B.14}$$

We define $\mathbf{T} = \mathbf{M}^{-1}\mathbf{B}^T(\mathbf{BM}^{-1}\mathbf{B}^T)^{-1}$. Multiplying Eq. (B.14) with $\mathbf{T}$ and inserting the expression in Eq. (B.12) leads to

$$\frac{d^2 r}{dt^2} = (\mathbb{1} - \mathbf{TB})\mathbf{M}^{-1}\boldsymbol{f} - \mathbf{T}\frac{d\mathbf{B}}{dt}\frac{dr}{dt}. \tag{B.15}$$

The matrix $\mathbb{1} - \mathbf{TB}$ acts as a projection operator to the $3N - K$ dimensional subspace to which the dynamics are confined by the constrains. To use Eq. (B.15) as an algorithm in MD simulation, the Taylor expansion, Eq. (3.3), is combined with Eq. (B.15)

$$r_{n+1} = 2r_n - r_{n-1} + [\mathbb{1} - \mathbf{T}_n\mathbf{B}_n](\Delta t)^2 \mathbf{M}^{-1}\boldsymbol{f} - \Delta t \mathbf{T}_n(\mathbf{B}_n - \mathbf{B}_{n-1})\boldsymbol{v}_{n-1/2}, \tag{B.16}$$

where the velocity is defined as

$$\boldsymbol{v}_{n-1/2} = \frac{r_n - r_{n-1}}{\Delta t} \tag{B.17}$$

As the constraints, Eq. (B.9), do not depend explicitly on time, the time derivative of the constraints yields

$$0 = \frac{d\boldsymbol{g}}{dt} = \sum_{j=1}^{3N} \frac{\partial \boldsymbol{g}}{\partial r_j}\frac{\partial r_j}{\partial t} = \mathbf{B}_n \boldsymbol{v}_{n+1/2}, \tag{B.18}$$

which allows Eq. (B.16) to be simplified to

$$r_{n+1} = r_n + (\mathbb{1} - \mathbf{T}_n\mathbf{B}_n)[r_n - r_{n-1} + (\Delta t)^2 \mathbf{M}^{-1}\boldsymbol{f}]. \tag{B.19}$$

The algorithm Eq. (B.19) is analytically correct, but numerically instable. As the constraints, Eq. (B.9), entered the algorithm only in the second time derivative, the algorithm accumulates numerical errors. An additional term $-\mathbf{T}_n(\mathbf{B}_n r_n - d)$, in which d denotes the vector of bond lengths, can solve this problem partially,

$$r_{n+1} = (\mathbb{1} - \mathbf{T}_n \mathbf{B}_n)[2 r_n - r_{n-1} + (\Delta t)^2 \mathbf{M}^{-1} f] + \mathbf{T}_n d. \tag{B.20}$$

Additionally the bond lengths must be modified, as Eq. (B.20) just removes the velocity components along the old bond direction but does not fix the value of the bond lengths. Again, this is a consequence of using the second time derivatives instead of the original constraints. If l_i is the length of the bond after an update, the projection of the new bond onto the old direction is set to $p_i = [2d_i^2 - l_i^2]^{1/2}$. Then, the position vector is

$$r_{n+1}^* = (\mathbb{1} - \mathbf{T}_n \mathbf{B}_n) r_{n+1} + \mathbf{T}_n p. \tag{B.21}$$

Eqs. (B.20) and (B.21) built the constrained version of the Verlet algorithm. The original Verlet algorithm, Eq. (3.5), is contained in the square brackets, but the matrix $(\mathbb{1} - \mathbf{T}_n \mathbf{B}_n)$ projects it such that it meets the constraints. An analogous Leap Frog version can be obtained [96].

The implementation of the constrained Verlet algorithm Eq. (B.20) involves the inverting of $\mathbf{B}_n \mathbf{M}^{-1} \mathbf{B}_n^T$ which is required for the calculation of $\mathbf{T}_n$. A proper rearrangement allows to write

$$(\mathbf{B}_n \mathbf{M}^{-1} \mathbf{B}_n^T)^{-1} - \mathbf{S}(\mathbb{1} \quad \mathbf{A}_n)^{-1} \mathbf{S}, \tag{B.22}$$

with a diagonal matrix $\mathbf{S}$. As $\mathbf{A}_n$ can be shown to be sparse, symmetric, and to have eigenvalues smaller than one, the Neumann series can be applied [224], i.e.,

$$(\mathbb{1} - \mathbf{A}_n)^{-1} = \sum_{k=0}^{\infty} \mathbf{A}_n^k. \tag{B.23}$$

Truncating the series in Eq. (B.23) after the forth term is a reasonable approximation and makes the inversion much more efficient [96, 97].

C Derivation of the Rouse ACF

The following treatment of the dynamics of the Rouse chain can be found in [149, 150, 158]. Here, as in the main text, the Rouse chain is treated as a one-dimensional object; the three-dimensional case is obtained by a superposition of the independent, one-dimensional time evolutions.

First assume an infinite chain. The modes of the chain can be expressed in terms of the Fourier modes. The solutions of the discrete eigenvalue equation

$$\tilde{\omega}^2[\Psi_\omega(i+1) - 2\Psi_\omega(i) + \Psi_\omega(i-1)] = -\omega^2 \Psi_\omega(i), \quad \text{(C.24)}$$

built an orthonormal basis set, $\sum_i \Psi_\omega(i)\Psi_{\omega'}(i) = \delta_{\omega\omega'}$. The Fourier modes are approximately given by

$$\Psi_\omega(i) = \cos\left(\frac{i\omega}{\tilde{\omega}}\right). \quad \text{(C.25)}$$

The positions can be expressed as a linear combination of the Fourier modes, Ψ_ω, i.e., $z_i(t) = \sum_{\{\omega\}} \zeta_\omega(t)\Psi_\omega(i)$. Essentially, this is the normal mode decomposition of the chain; $\zeta_\omega(t)$ are the normal modes. For a finite chain, the forces at the ends of the chain equal zero. As a consequence, the set of possible eigenfrequencies, $\{\omega_n\}$, is discrete, finite, and obeys Eq. (5.12),

$$\omega_n = \frac{\tilde{\omega}\pi n}{N-1} \quad \text{with } n = 0, 1, ..., N-1. \quad \text{(C.26)}$$

We use the notation $\zeta_n = \zeta_{\omega_n}$ for the normal modes.

We now derive the autocorrelation of the distance between bead i and j. The distance between bead i and j is

$$\Delta(t) = z_i(t) - z_j(t) \quad \text{(C.27)}$$

$$= \sum_n \zeta_n(t)[\Psi_n(i) - \Psi_n(j)] \quad \text{(C.28)}$$

$$= \sum_n \zeta_n(t)\left[\cos\left(\frac{i\omega_n}{\tilde{\omega}}\right) - \cos\left(\frac{j\omega_n}{\tilde{\omega}}\right)\right]. \quad \text{(C.29)}$$

The identity

$$\cos\alpha - \cos\beta = 2\sin\left(\frac{\alpha+\beta}{2}\right)\sin\left(\frac{\beta-\alpha}{2}\right), \quad \text{(C.30)}$$

allows the deviation, $\Delta(t)$, to be expressed as

$$\Delta(t) = 2\sum_n \zeta_n(t) \sin\left(\frac{(i+j)\omega_n}{\tilde{\omega}}\right) \sin\left(\frac{(j-i)\omega_n}{\tilde{\omega}}\right) \quad \text{(C.31)}$$

$$\approx 2\sum_n \zeta_n(t) \frac{(j-i)\omega_n}{\tilde{\omega}}. \quad \text{(C.32)}$$

The approximation in the last line can be performed if the deviation depends only on the distance of the beads on the chain, *i.e.*, as long as $i - j \ll N$ and both beads are not at the ends of the chain. Then, the ACF is

$$\langle \Delta(t+\tau)\Delta(\tau)\rangle_\tau = \sum_{nk} \langle \zeta_n(t+\tau)\zeta_k(\tau)\rangle_\tau \frac{\omega_n \omega_k}{\tilde{\omega}^2}(j-i) \qquad (\text{C.33})$$

$$= \sum_n \frac{k_B T}{m\tilde{\omega}^2} e^{-tn^2/\tau_R}(j-i), \qquad (\text{C.34})$$

where the vanishing correlation between different normal modes is used, as follows from the equipartition theorem, Eq. (5.15). For long chains, the sum can be replaced by an integral. Hence,

$$\langle \Delta(t+\tau)\Delta(\tau)\rangle_\tau = \int_0^\infty \frac{k_B T}{m\tilde{\omega}^2} e^{-tn^2/\tau_R}(j-i)\,\mathrm{d}n \qquad (\text{C.35})$$

$$= \frac{k_B T}{m\tilde{\omega}^2}(j-i)\sqrt{\frac{\tau_R}{t}} \int_0^\infty e^{-y^2}\,\mathrm{d}y. \qquad (\text{C.36})$$

Therefore, the time dependence of the ACF is

$$\langle \Delta(t+\tau)\Delta(\tau)\rangle_\tau \propto t^{-1/2}. \qquad (\text{C.37})$$

D Fractional diffusion equation

In Sec. 5.3, the fractional diffusion equation (FDE) is employed, serving as the macroscopic equation governing the time evolution of an ensemble of CTRW processes. The equation can be obtained as the continuum limit of the CTRW with a power-law WTD as given in Eq. (5.20). In this section, the derivation of the FDE is reviewed, using Fourier-Laplace transforms, as is done in Ref. [8]. First, some mathematical notations are introduced.

D.1 Gamma function

The Riemannian Gamma function is defined as

$$\Gamma(\alpha) = \int_0^\infty t^{\alpha-1} e^{-t}\,\mathrm{d}t. \qquad (\text{D.38})$$

It obeys the functional equation

$$\Gamma(\alpha+1) = \alpha\Gamma(\alpha). \tag{D.39}$$

The latter can be used as a recurrence relation. As $\Gamma(1) = 1$, if follows for $n \in \mathbb{N}$

$$\Gamma(n+1) = n!. \tag{D.40}$$

Therefore, the Gamma function is a continuous extension of the factorial.

Further, a useful property of the Gamma function is

$$\Gamma(1+\alpha)\Gamma(1-\alpha) = \frac{\pi\alpha}{\sin\pi\alpha}, \qquad \text{with } \alpha \in \mathbb{R} \setminus \mathbb{Z}. \tag{D.41}$$

The incomplete Gamma function is defined as

$$\Gamma(\alpha, x) = \int_x^\infty t^{\alpha-1} e^{-t} dt. \tag{D.42}$$

The incomplete Gamma function $\Gamma(-\alpha, u)$ can be expanded as a chain fraction [107]. It has the following asymptotic behavior for small x

$$\Gamma(\alpha, x) \approx \Gamma(\alpha) - \frac{e^{-x} x^\alpha}{\alpha} \qquad \text{for } x \ll 1. \tag{D.43}$$

D.2 Fractional derivatives

Let $_0J_t$ be the integral operator defined by

$$_0J_t \phi(t) = \int_0^t \phi(t') dt'. \tag{D.44}$$

According to Cauchy's formula for repeated integrals, the n-fold integral, n being integer, can be expressed as

$$_0J_t^n \phi(t) = \frac{1}{(n-1)!} \int_0^t \frac{\phi(t')}{(t-t')^{1-n}} dt'. \tag{D.45}$$

A straightforward way to extend Eq. (D.45) from $n \in \mathbb{N}$ to real $\alpha > 0$, is to define the integral operator

$$_0J_t^\alpha \phi(t) = \frac{1}{\Gamma(\alpha)} \int_0^t \frac{\phi(t')}{(t-t')^{1-\alpha}} dt'. \tag{D.46}$$

D Fractional diffusion equation

Let $\beta = n - \alpha$, with $0 < \alpha < 1$ and $n \in \mathbb{N}$. The fractional derivative of order $\beta \in \mathbb{R}$ is defined as

$$_0\mathcal{D}_t^\beta \phi(t) = \frac{\mathrm{d}^n}{\mathrm{d}t^n} {}_0J_t^\alpha \phi(t). \tag{D.47}$$

The operator $_0\mathcal{D}_t^\beta$ is called the Riemann-Liouville operator. The fractional derivative of order $\alpha \neq 0$, of a power of t is given as

$$_0\mathcal{D}_t^{1-\alpha} t^\gamma = \frac{\Gamma(1+\gamma)}{\Gamma(\gamma+\alpha)} t^{\gamma-1+\alpha}. \tag{D.48}$$

The Laplace transform of the Riemann-Liouville operator reads

$$\mathcal{L}[_0\mathcal{D}_t^{-\beta}\phi(t)] = u^{-\beta}\mathcal{L}[\phi(t)]. \tag{D.49}$$

D.3 The fractional diffusion equation

Let a CTRW have the WTD $w(t) \sim (t/\tau_0)^{-(1+\alpha)}$, as in Eq. (5.20), with an WTD exponent $0 < \alpha < 1$. The JLD, $\varphi(x)$, is assumed to be symmetric around $x = 0$ with variance $\bar{x}^2$. The probability of being at position x at time t is denoted as $W(x,t)$. In this subsection, the equation is derived that governs the time evolution of the probability distribution $W(x,t)$. The initial condition is $W_0(x) = \delta(x)$, i.e., the CTRW starts at $t = 0$ at the position $x = 0$.

The probability of having just arrived at position x at time t, denoted by $\eta(x,t)$, equals the probability of having arrived at x' at time t' *and* of waiting a time $t - t'$ at x' *and* to make a jump of length $x - x'$ integrated over all possible x' and t'. Mathematically this equality is expressed by

$$\eta(x,t) = \int_{-\infty}^{\infty} \int_0^t \eta(x',t') w(t-t') \varphi(x-x') \mathrm{d}t' \mathrm{d}x' + W_0(x)\delta(t). \tag{D.50}$$

The probability $W(x,t)$ can be obtained from $\eta(x,t)$ as

$$W(x,t) = \int_0^t \eta(x,t') \left[1 - \int_0^{t-t'} w(t'') \mathrm{d}t'' \right] \mathrm{d}t'. \tag{D.51}$$

The convolution equations take a relatively simple form in the Fourier-Laplace representation, in which the variables x and t are transformed to k (Fourier transform[1]) and u (Laplace transform[2]), respectively. Combining the Fourier-Laplace transforms of Eqs. (D.50) and (D.51), the following equation is obtained

$$W(k,u) = \frac{1-w(u)}{u}\frac{W_0(k)}{1-\varphi(k)w(u)}. \tag{D.54}$$

The Fourier transform of the JLD is for small k approximately given as

$$\varphi(k) \approx 1 - \bar{x}^2 k^2 + \mathcal{O}(|x|^4). \tag{D.55}$$

The Laplace transform, $t \to u$, of the WTD with power-law tail, [Eq. (5.20)], $w(u) = \alpha e^u u^\alpha \Gamma(-\alpha, u)$, using the incomplete gamma function. With the asymptotic behavior of $\Gamma(\alpha, u)$ in Eq. (D.43), the Laplace transform $w(u)$ is derived

$$w(u) \approx 1 - \Gamma(1-\alpha)(\tau_0 u)^\alpha. \tag{D.56}$$

With the generalized diffusion constant, $K_\alpha = \bar{x}^2/[2\Gamma(1-\alpha)\tau_0^\alpha]$, the diffusion limit, $\bar{x}^2 \to 0$, $\tau_0 \to 0$, and $K_\alpha = \text{const.}$, can be performed. The diffusion limit corresponds to the range of large x- and t-values, $i.e.$, the time, t, is large compared to the typical time scale, $t \gg \tau_0$, and the typical x is large compared to the mean squared jump length, $x \gg \sqrt{\bar{x}^2}$. In the Fourier-Laplace space, the diffusion limit is performed by the limit $(k, u) \to 0$. Inserting the Eqs. (D.55) and (D.56) in Eq. (D.54), a rearrangement of the terms yields the following equation

$$W(k,u) - W_0(k)/u = -K_\alpha k^2 u^{-\alpha} W(k,u). \tag{D.57}$$

In the original coordinates using Eq. (D.49), and with an additional time derivative, the FDE follows[3]

$$\frac{\partial}{\partial t} W(x,t) = {_0\mathcal{D}_t^{1-\alpha}} K_\alpha \frac{\partial^2}{\partial x^2} W(x,t). \tag{D.58}$$

---

[1] The Fourier transform of a function, $\phi(x)$ is given as

$$\phi(k) = \int_{-\infty}^{\infty} \phi(x) e^{-ikx} dx. \tag{D.52}$$

[2] The Laplace transform of a function, $\psi(t)$ is given as

$$\psi(u) = \int_0^\infty \psi(t) e^{-ut} dt. \tag{D.53}$$

[3] If the Laplace representation of a WTD with finite mean value is substituted in Eq. (D.54),

D Fractional diffusion equation

D.4 The Mittag-Leffler function

The Mittag-Leffler function (MLF) is defined as

$$\mathrm{E}_\alpha(z) = \sum_{n=0}^{\infty} \frac{z^n}{\Gamma(1+n\alpha)}, \qquad (\text{D.59})$$

where $\alpha > 0$. For $\alpha = 1$, the MLF equals the exponential function, for $\alpha = 1/2$, it can be written in terms of the error function, $\mathrm{E}_{1/2}(z) = e^{-z^2}[1+\mathrm{erf}(z)]$. Using Eq. (D.49), the Laplace transform of the MLF reads

$$\mathcal{L}[\mathrm{E}_\alpha(-cz^\alpha)] = \sum_{n=0}^{\infty} \frac{1}{\Gamma(1+n\alpha)} \frac{c^n \Gamma(\alpha n + 1)}{u^{\alpha n + 1}} = \frac{1}{u + c u^{1-\alpha}}. \qquad (\text{D.60})$$

The MLF, $\mathrm{E}_\alpha(-[t/t_m]^\alpha)$, has an asymptotic behavior given as

$$\mathrm{E}_\alpha(-[t/t_m]^\alpha) \approx \begin{cases} \exp\left(-\frac{[t/t_m]^\alpha}{\Gamma(1+\alpha)}\right) & t \ll t_m \\ \frac{t_m^\alpha}{t^\alpha \Gamma(1-\alpha)} & t \gg t_m \end{cases}. \qquad (\text{D.61})$$

Therefore, the MLF has a stretched exponential behavior for small t, whereas the large-t behavior is given by a power law.

Let $0 < \alpha < 1$, and apply the Riemann-Liouville operator to the definition of the MLF, using Eq. (D.48)

$$_0\mathcal{D}_t^{1-\alpha} \mathrm{E}_\alpha(-ct^\alpha) = \sum_{n=0}^{\infty} \frac{(-c)^n}{\Gamma(1+n\alpha)} \frac{\Gamma(\alpha n + 1)}{\Gamma(\alpha n + \alpha)} t^{\alpha n - 1 + \alpha} \qquad (\text{D.62})$$

$$= \sum_{n=1}^{\infty} \frac{(-c)^{n-1}}{\Gamma(n\alpha)} t^{\alpha n - 1} \qquad (\text{D.63})$$

$$= -\frac{1}{c} \sum_{n=1}^{\infty} \frac{(-c)^n}{\Gamma(1+n\alpha)} \alpha n t^{\alpha n - 1} \qquad (\text{D.64})$$

$$= -\frac{1}{c} \frac{\partial}{\partial t} \mathrm{E}_\alpha(-ct^\alpha). \qquad (\text{D.65})$$

---

the derivation leads to the classical diffusion equation, Eq. (2.33). As only the first terms in the small-u and small-k expansion determine the diffusion limit, the particular analytical form of WTD and JLD are not significant for the diffusion equation to be valid. The diffusion process is characterized by the variance of the JLD and the mean waiting time (the mean of the WTD). The anomalous diffusion seen for a WTD with power-law tail, Eq. (5.20), appears as a consequence of the diverging mean value of the WTD.

Therefore the function
$$a_k(t) = E_\alpha\left(-K_\alpha k^2 t^\alpha\right) \tag{D.66}$$
is the solution of the equation
$$\frac{\partial}{\partial t} a_k(t) = -K_\alpha k^2 \, {}_0\mathcal{D}_t^{1-\alpha} a_k(t). \tag{D.67}$$

This allows the Fourier decomposition to be applied, analogously to the classical diffusion equation. The prominent role of the MLF in the context of fractional differential equations roots in its correspondence to the exponential function. For the results of classical diffusion problems to be extended to fractional diffusion, it suffices in many cases to replace the exponential time dependence by an MLF time dependence.

E Exponentials and power laws

The sum of exponentials can give rise to a power-law behavior, as in Eq. (5.94). Here, we present a brief, non-rigorous argument demonstrating under which conditions the sum of negative exponentials yields a power law.

Consider the following function
$$f(t) = 1 - R\exp(-\mu t), \tag{E.68}$$

which is characterized by two parameters, R and μ. The function $f(t)$ is illustrated in Fig. E.1 with $\mu = 1$ for various R. The function $f(t)$ has the limiting cases $f(0) = 1-R$ and $f(t \to \infty) \to 1$. In Fig. E.1 it is apparent that $f(t)$ exhibits a power-law like transition between the two limiting cases, $i.e.$, in an intermediate range of t, the function $f(t)$ is very similar to a power law. More precisely, in the range $t \in [-\log(0.3)/\mu; -\log(0.7)/\mu]$, where $0.3 \leqslant \exp(-\mu t) \leqslant 0.7$, the function $f(t)$ can be approximated as a power law, $f(t) \approx ct^{\alpha(R)}$. Approximately, the following exponents are seen

$$\alpha(R) = \log\left(\frac{1-0.3R}{1-0.7R}\right) / \log\left(\frac{\log 0.3}{\log 0.7}\right). \tag{E.69}$$

E Exponentials and power laws

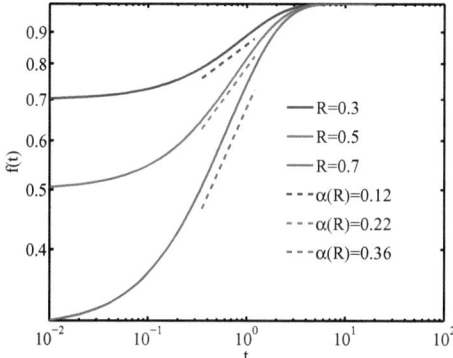

Figure E.1: *(Color online)* **Function $f(t)$ of Eq. (E.68)** - The function $f(t)$ of Eq. (E.68) has the value $1 - R$ for $t = 0$ and 1 for $t \to \infty$. The intermediate behavior around the time scale μ^{-1} can be approximated by a power-law curve. $f(t)$ is illustrated for various R, see legend. The approximate power-law exponents are also given in the legend.

A finite series of terms like Eq. (E.68) can lead to a power law over an arbitrary interval $[t_1, t_2]$, i.e.

$$1 - \sum_k R_k \exp(-\mu_k t) \sim t^\alpha, \text{ for } t \in [t_1, t_2] \quad (E.70)$$

as in Eq. (5.94). The power law at t^* is dominated by terms with

$$\mu_k \approx \log(2)/t^*. \quad (E.71)$$

For simplicity, we assume the μ_k to be so different that only one of the μ_k obeys Eq. (E.71) at a given t^*, i.e., if $\mu_k = \log(2)/t^*$ then $\exp(-\mu_j t^*) \approx 1$ or $\exp(-\mu_j t^*) \approx 0$ for all $j \neq k$. Hence, we assume that for any given time scale only one R_k-term must be taken into account.

In the general case, various R_k contribute at a given time t^*, i.e., various eigenvectors can contribute to the same time scale. The present argument treats them as if their respective transport coefficients were summed to one common R_k. At the times $t_k = \log(2)/\mu_k$ the values of the left side of Eq. (E.70) shall be on the curve ct^α. Therefore, it is

$$c\mu_k^{-\alpha}[\log 2]^\alpha = \sum_{i<k} R_i + \frac{R_k}{2}. \tag{E.72}$$

A further condition can be imposed,

$$R_2 = \frac{R_k}{\sum_{i<k} R_i}. \tag{E.73}$$

The above equation represents the fact that the weight of the slowest term, R_2, and the long-time plateau, R_1, have the same ratio as the weight of the term k, dominating at time $\tau_k = \mu_k^{-1}$, and the slower contributions, $\sum_{i<k} R_i$.

The Eqs. (E.72) and (E.73) characterize the conditions under which a sum of exponentials like in Eq. (E.70) gives rise to an power-law behavior on a certain, intermediate interval. Note that Eq. (5.94) has the same form as Eq. (E.70). In Fig. 5.18, it is illustrated how a small number of transport coefficients sum up to the power-law MSD in the case of the Sierpiński triangle. Employing the Laplace representation allows a continuous distribution of transport coefficients equivalent to Eq. (E.72) to be established.

Zitate in den Kapitelüberschriften

- THOMAS S. KUHN: The Structure of Scientific Revolutions. – Chicago (IL): The University of Chicago Press 1973, 2nd edition, S. 24

- HERMANN VON HELMHOLTZ: Antwortrede gehalten beim Empfang der Graefe-Medaille zu Heidelberg, am 9. August 1886 in der Aula der Universität zu Heidelberg. In: Vorträge und Reden. – Band 2, Braunschweig: F. Vieweg und Sohn 1903, S. 318

- FRIEDRICH SCHILLER: Wallensteins Tod, II, 2.

- D. FRENKEL & B. SMIT: [18], S. 73.

- RICHARD P. FEYNMAN: The Feynman lectures, vol. 1, 5th edition, 1970, S. 3-6.

- NICOLAUS VON KUES: Trialogus de possest, ca. 1460, S. 44. Zitiert nach: http://www.hs-augsburg.de/~harsch/Chronologia/Lspost15/Cusa/cus_tria.html.

- DAVID RUELLE: Hasard et Chaos. – Editions odile jacob 2000, S. 116.

- GOTTFRIED WILHELM FRIEDRICH LEIBNIZ: Brief an Guillaume François Antoine Marquis de l'Hospital vom 30. September 1695.

I want morebooks!

Buy your books fast and straightforward online - at one of the world's fastest growing online book stores! Environmentally sound due to Print-on-Demand technologies.

Buy your books online at
www.get-morebooks.com

Kaufen Sie Ihre Bücher schnell und unkompliziert online – auf einer der am schnellsten wachsenden Buchhandelsplattformen weltweit!
Dank Print-On-Demand umwelt- und ressourcenschonend produziert.

Bücher schneller online kaufen
www.morebooks.de

OmniScriptum Marketing DEU GmbH
Heinrich-Böcking-Str. 6-8
D - 66121 Saarbrücken
Telefax: +49 681 93 81 567-9

info@omniscriptum.com
www.omniscriptum.com

Printed by Books on Demand GmbH, Norderstedt / Germany